Plant Breeding for Abiotic Stress Tolerance

Roberto Fritsche-Neto · Aluízio Borém
Editors

Plant Breeding for Abiotic Stress Tolerance

Springer

Editors
Roberto Fritsche-Neto
Departamento de Fitotecnia
Universidade Federal de Viçosa
Av. PH Rolfs s/n
Viçosa, MG 36570-000
Brazil

Aluízio Borém
Departamento de Fitotecnia
Universidade Federal de Viçosa
Av. PH Rolfs s/n
Viçosa, MG 36570-000
Brazil

ISBN 978-3-642-30552-8 ISBN 978-3-642-30553-5 (eBook)
DOI 10.1007/978-3-642-30553-5
Springer Heidelberg New York Dordrecht London

Library of Congress Control Number: 2012940243

Printed on acid-free paper

Springer is part of Springer Science+Business Media (www.springer.com)

Preface

Practical experience shows that abiotic stresses occur at high or low intensity in just about all agricultural growing areas around the globe. In some regions the stress due to heat, drought, and mineral nutrition or even due to aluminum toxicity may not be present in a specific year, but this could change in the coming years. The consequences of the climate change for agriculture have caused concern for many policy makers around the world and it has been a hot topic in many scientific forums. The main apprehension is that the world may not be able to sustain its food production under the crescent abiotic stresses with global warming.

The organizers of this book have pulled together this publication aiming to cluster the most relevant scientific achievements and the state of the art in plant breeding and cultivars development for abiotic stresses.

Written by experts in different areas of abiotic stresses, in an easy to understand language, this book is an obligated reference for all interested in plant breeding and in the upcoming challenges that agriculture will face with climate change.

Roberto Fritsche-Neto
Aluízio Borém

Contents

1 **Abiotic Stresses: Challenges for Plant Breeding in the Coming Decades** 1
Aluízio Borém, Magno Antonio Patto Ramalho and Roberto Fritsche-Neto

2 **Breeding for Stress-Tolerance or Resource-Use Efficiency?** 13
Roberto Fritsche-Neto and Júlio César DoVale

3 **The Physiology of Abiotic Stresses** 21
Paulo C. Cavatte, Samuel C. V. Martins, Leandro E. Morais, Paulo E. M. Silva and Fábio M. DaMatta

4 **Breeding for Nitrogen Use Efficiency** 53
Júlio César DoVale, Rodrigo Oliveira DeLima and Roberto Fritsche-Neto

5 **Breeding for Phosphorus Use Efficiency** 67
Sidney Netto Parentoni, Flávia Ferreira Mendes and Lauro José Moreira Guimarães

6 **Breeding for Water Use Efficiency** 87
Marcelo de Almeida Silva, Claudiana Moura dos Santos, Carlos Alberto Labate, Simone Guidetti-Gonzalez, Janaina de Santana Borges, Leonardo Cesar Ferreira, Rodrigo Oliveira DeLima and Roberto Fritsche-Neto

7 **Breeding for Salinity Tolerance** 103
Nand Kumar Fageria, Luís Fernando Stone and Alberto Baêta dos Santos

8 Breeding for Aluminum Tolerance 123
Lauro José Moreira Guimarães, Claudia Teixeira Guimarães,
Jurandir Vieira Magalhães, Sidney Netto Parentoni
and Flávia Ferreira Mendes

9 Breeding for Heat-Stress Tolerance. 137
Moacil Alves de Souza, Adérico Júnior Badaró Pimentel
and Guilherme Ribeiro

10 Breeding Perennial Species for Abiotic Stress 157
Rinaldo Cesar de Paula, Nadia Figueiredo de Paula
and Celso Luis Marino

Index 173

Chapter 1
Abiotic Stresses: Challenges for Plant Breeding in the Coming Decades

Aluízio Borém, Magno Antonio Patto Ramalho and Roberto Fritsche-Neto

Abstract Modern agriculture has been providing food, feed, fiber and more recently biofuel to meet the Worlds demand. One of the bases of this modern agriculture is the improved cultivars, which are much higher yielder than ancient ones. The scientific literature documents the significant contribution of plant breeding to agriculture and finally food production worldwide. Now-a-days, agriculture has new and huge challenges, due to population growth, the pressure on agriculture liability on the environmental conservation, and climate change. To cope with these new challenges, many plant breeding programs have reoriented their breeding scope to stress tolerance in the last years. So, in this book, experts on plant physiology and on plant breeding presents the most recent advances and discoveries applied to abiotic stresses, discussing the new physiological concepts, breeding methods, and modern molecular biological approaches to develop improved cultivars tolerant to most sorts of abiotic stress.

Keywords Climate change · Global warming · Improved cultivars · Breeding for draught · Biotechnology

A. Borém (✉) · R. Fritsche-Neto
Universidade Federal de Viçosa, Viçosa, Brazil
e-mail: borem@ufv.br

R. Fritsche-Neto
e-mail: roberto.neto@ufv.br

M. A. P. Ramalho
Universidade Federal de Lavras, Lavras, Brazil
e-mail: magnoapr@dbi.ufla.br

R. Fritsche-Neto and A. Borém (eds.), *Plant Breeding for Abiotic Stress Tolerance*,
DOI: 10.1007/978-3-642-30553-5_1, © Springer-Verlag Berlin Heidelberg 2012

1.1 Introduction

Agriculture was the most important event in the man's history, changing his nomadic lifestyle to start to settle in communities. Archeological evidences suggest that agriculture began about 10,000–12,000 years ago. With agriculture, man accelerated his influence in the environment, especially by substituting the native and diverse flora by the species of their preference or needs.

At the beginning of the nineteenth century, there was an increasing concern that the rapid growth of the population could result food shortage and the poor-resource population would go famine. The most famous writing about this was by the economist Thomas Malthus, who said that the population increase according to a geometric progression while the food supply would increase according to an arithmetic progression. While Malthus model was correct, he did consider the agricultural technology development such as use of fertilizers and especially the adoption of genetically improved cultivars. Therefore, even with the population increasing to 4.68 billion people during the twentieth century, food was available to meet the demand, except for short periods of time. It is important to say that today hunger is mainly due to wealth distribution and not due to food scarcity.

With modern agriculture using improved cultivars and good crop management, crop yield has been raised in just all agronomic species. This progress was more evident after the Second World War, with the wide adoption of improved cultivars, fertilization, insecticides, and fungicides. With the success of modern agriculture to produce food in quantity and quality, the prices of food dropped during the twentieth century. The food abundance, as a result of agricultural technology, had its own toll; society lost the sense that agriculture had importance. With plenty of food at low prices, many other areas of the human knowledge start to drain a significant part of the investments in R&D, and today the agricultural research lives a huge dilemma; how to develop the technology for the coming decades, especially with the new challenges with climate change. One of the signs of the shortage of investments for agricultural research is the small number of newly trained plant breeders in the last decade, in spite of plant breeding's contribution to increase food production worldwide (Vencovsky and Ramalho 2006; Duvick et al. 2004).

Although the population that goes hungry has been reduced in the last 50 years, there are evidences that it could spike in the medium and long run. Currently, about 925 million people do not have access to the minimum daily amount of food recommended by United Nations WHO considering the macronutrients (carbohydrates, proteins, and lipids). It is estimated that another 1 billion people have the occult hunger, that is, have bad nutrition with deficiency of vitamins and minerals (Evans, 1998; Borém and Rios 2011). On the other hand, nearly another 1 billion people is consuming food in excess with risk of obesity and diabetes.

Nowadays, agriculture has new and huge challenges, which tend to be even bigger in coming years. This is due to the fact that the population is supposed to increase up to 2050, when according to the United Nation FAO the world population will stabilize reaching 9.1 billion inhabitants. The population increase will have profound

Table 1.1 World food and fiber production demand in 2005 and 2025 (in millions of tons)

Product	Production in 2005	Demand in 2025	Additional production
Grains	2.219,40	3.140,40	921
Oily grains	595,05	750,97	155,96
Perennial crops	242,81	321,99	70,18
Annual crops	352,2	437,98	85,78
Coffee	7,72	9,4	1,68
Fibers	28,5	36,37	7,87
Timber	3.401,90	4.148,40	746,5

Source Adapted from Agroanalysis (2007)

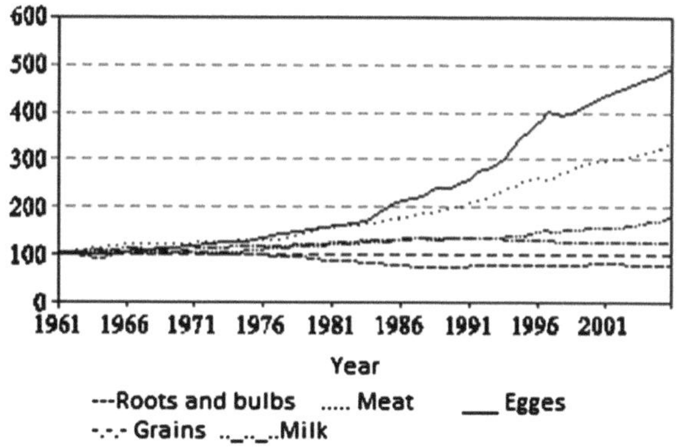

Fig. 1.1 *Per capita* consumption of five food products in emerging countries from 1996 to 2001. *Source* Agroanalysis (2011)

outcomes to our planet. It is surprising that, even before this sentence is complete, there will be another three inhabitants on the Earth, that is, a new person is born every three seconds, 259,000 new inhabitants per day, 7 million per month. It is worth noting that it took several million years for the Earth to reach its first 2 billion people, and that another 2 billion will be brought just in the next 25 years. Additionally, people are living longer and also moving from the country side to urban areas. Half of the world population currently live in cities and small towns, but this number should increase up to 60 % by 2030, when it is supposed to exist 26 cities with population over 10 million people (Beddington 2010).

The challenges to be faced by man due to the population increase in the coming decades are various and often debated (UK Government 2011). One of them is how to provide food to meet the demand. On Table 1.1 is listed the world food production and demand in 2005 and in 2025. The increased demand for food is not only due to population increase, but also due to many other factors, among them is the increase in population wealth, especially in the emerging countries called BRICS (Brazil, Russia, India, China, and South Africa), resulting in demand for a more diverse and rich diet (Fao 2009, 2010; Fig. 1.1a). It is well known that an increase in the per capita income results in an increase in food demand, and the BRICS countries represent 40 % of the world population and nearly 18 % world economy.

Another factor pressing the food production is the competition between the areas designed for bioenergy production. The demand for bioenergy should increase 45 % in the next 25 years, and growing areas used for grain will be allocated to sugarcane, jatropha, and other energy crops (Mitchel 2008; Beddington 2010).

The pressure on agriculture will be even greater due to its environmental liability. Among the environmental aspects used as input in agricultural water deserves a special attention. It is estimated that 70 % of the drinkable water is currently used for irrigation (Chris and Briscoe 2001). Therefore, water must be used more efficiently. Additionally, it is estimated that many areas around the world will have less soil water, creating the need for irrigation to be more widespread and intense in many growing areas.

The renewable agriculture inputs will also become scarcer and more costly. For example, Brazil uses 2.5 millions of nitrogen, i.e., 2.5 % of the world consumption, and 3.7 million of K_2O, making up to 13.9 % of world consumption.

To complicate the scenario even more, there is an increasing pressure for lowering the CO_2 emission (Wright 2010) and it is regarded agriculture will suffer most (Lobell et al. 2003; Assad and Pelegrino 2007). To cope with this scenario many marginal areas will be added to the agricultural system. Many of those marginal areas have aluminum toxicity, salinity, or other abiotic stresses, demanding for cultivars adapted to such constrains.

Most of the agricultural systems are not sustainable, mainly due to the intense use of non-renewable resources reinforcing the feeling for more stress-tolerant cultivars.

The lack of sustainability will make the insects, diseases, weeds, and biotic stresses more prevalent and severe. To deal with those challenges, more planning and new technology is necessary. Besides the traditional techniques and breeding methods, Biotechnology (Quam 2009; Borém and Almeida 2011) and other emerging technologies, such as Nanotechnology (Alexandratos 2006; Bruinsma 2009) must be adopted.

The contribution of plant breeding to increase food, feed, fiber, and biofuel production (Vencovsky and Ramalho 2006; Duvick et al. 2004) is well documented. Nevertheless, whatever happens in the coming decade, plant breeders will have to devote profound dedication and deep scientific knowledge to keep pace with the increasing food demand in a hungry world. The main goal of this chapter is to address the scope of abiotic stresses for agriculture and how to cope with them, helping plant breeders to establish right priorities and develop adequate cultivars.

1.2 Crop Yield Potential

Before discussing what potential yield is, it is needed to define biologic efficiency. The concept of biologic efficiency varies with the crop species and situations. For example, for food, feed, and fiber biologic efficiency depends on the solar energy conversion. Yield should reflect the unit of the product per unit of absorbed solar energy. Crop scientists and economists, among others, prefer to associate biologic efficiency to unit of area, such as amount of the product (kg, tons) per area unit

Table 1.2 Record grain yield (ton.ha^{-1}) in different crop species

Species	Productivity	Source
Rice	10.5 (15.9)[a]	Boyer (1987)
Corn	23.2	Dwick and Cassman (1999)
Wheat	14.1	Tollenaar and Lee (2002)
Sorghum	20.1	Ort and Long (2003)
Soybean	7.3 (22.5)[b]	Ort and Long (2003)

[a] Yield estimated in function of solar radiation during the growing season (Peng et al.1999)
[b] Yield estimated in function of photosynthetic efficiency (Specht et al. 1999)

(acre, hectare). In this book and in most situations for plant breeders, the biologic efficiency is measured by amount of the product per unit of applied resources. In many cases, it is estimated per unit of the following resources land, water, energy, fertilizer, pesticides, capital, and machinery. The efficiency is, therefore, relative and varies with the environment and the cultivar. Thus, the use of a biologic index can help plant breeders to develop cultivars with higher efficiency of the use of natural resources.

Yield potential is the maximum production when the cultivar is grown in a favorable environment, that is, in the absence of limiting factors and where it is completely adapted. In this environment, mineral nutrients and water are not yield-limiting factors and pests and weeds are effectively controlled, i.e., the plant development takes place with no stress (Evans and Fisher, Evans and Fisher 1999). In general, it is difficult or even impossible to grow a crop under conditions to obtain the maximum yield. However, there are several reports on the literature on maximum yield for many crop species. There are also reports on potential yield in function of the solar radiation. All those estimates are much higher than those estimates obtained in practice (Table 1.2). The yield potential is substantially higher than the biologic yield per land unit, i.e., yield in ton.ha^{-1}, a unit often used by agronomists. Therefore, the cultivar genetic potential is much higher than that obtained in real life. Why does that occur? As previously stated, each environmental factor that negatively affects the plant development is called *stress*. There are many environmental and/or management factors that have profound impact on plant growth, and the restriction of one or more of those factors diminishes the biologic efficiency with reflection on the productivity.

The genetic gain for most crops, if not all, obtained in the last century, was due to plant selection targeting tolerance toward stresses, instead of aiming higher yield potential (Tollenaar and Wu 1999; Tollenaar and Lee 2002). It is believed that the same strategy will continue to be used for breeding in near future.

Boyer (1982) reports the main causes of crop failure in the USA. It can be observed that the abiotic stresses are responsible for up to 89 % of all crop failure, and among them drought is responsible for more than 40 % of the losses (Table 1.3).

It must be realized that stresses in tropical growing areas tend to be much more prevalent and intense than in temperate regions. Paterniani (1990) draws a parallel between environments' growing factors in those regions where corn is grown (Table 1.4). It is noteworthy that stresses for agriculture are substantially more

Table 1.3 The main reasons for crop loses, in percentages, where insurance was granted in the USA

Reason of loss	Loss (%)
Drought	40.8
Flooding	18.2
Cold	13.8
Hail	11.3
Wind	7.0
Insects	4.5
Diseases	2.7
Others	1.5

Source Adapted from Ort and Long (2003)

Table 1.4 Characteristics of corn grown in tropical and temperate regions

Characteristics	Temperate	Tropical
Annual Climatic variation	Relatively stable	Unpredictable
Annual rain variation	Relatively uniform	Variable
Rain among regions	Relatively uniform	Variable
Photoperiod	Long days	Short days
Night Temperature	Cold	Warm
Soil conditions	Usually favorable	Usually adverse
Sowing period	Restrict/well defined	Highly variable/all year long
Growing period	Well defined	Highly variable
Germination conditions	Cold soil/soil fungi present	Soil insects present
Weed infection	Medium infestation	High infestation
Grain insects	Low infestation	High infestation starting in the field

Source Adapted from Paterniani (1990)

adverse in tropical regions than in temperate regions. The author compares, for example, the rain precipitation in Piracicaba, SP (Brazil) and in Ames. IA (USA). In Ames, the rain in July and August, months when the crop flowers, averaged 92.6 mm ± 35.5 and 97.6 mm ± 40.7, respectively. In Piracicaba, during the months when the crop is flowering, i.e., December and January, the average rain is 218.9 mm ± 75.9 and 216 mm ± 94.1, respectively. At first, those estimates suggest that water availability is much more favorable in Piracicaba. However, the rain distribution in Piracicaba is much more erratic. In this region, short periods of intense rain occur with low accumulation of the water in the soil. Those periods of heavy rain usually are followed by long periods with no precipitation.

The author did not consider the environmental factors that may be controlled by farmers, such as mineral nutrition. Thus, it is worth noting the importance of abiotic stresses in the region in the Globe where food production is expanding at higher rate.

The main objective of plant breeding is to develop improved cultivars. Usually, they are developed with more drought tolerance or with resistance to any other stress, but breeders are now also striving to develop cultivars more efficient in using limiting environmental factors.

1.3 Plant Breeding and Stresses Caused by Climate Change

Practical experience shows that abiotic stresses occur at high or low intensity in just about all agricultural growing areas around the Globe. In some regions, the stress due to drought, heat, and mineral nutrition or even due to aluminum toxicity may not be present in a specific year, but this could change in the coming years. The consequences of the climate change for agriculture have caused concern for many policy makers and been a hot topic and many scientific forums (Ramalho et al. 2009).The main apprehension is that the world may lack ability to sustain its food production under the increasing abiotic stresses with Global Warming.

Climate change is expected to bring additional challenges agriculture, already under pressure for higher productions to satisfy the crescent demand for food. Climate change will affect the development of new crop cultivars. Heavy rains, long droughts, intense heat will become more frequent and extreme, increasing volatility of crop production and food security. Food production is expected to be also affected by sea currents and ocean temperature. One could initially think that the increase of atmospheric CO_2, the raw matter for plants to grow, could foster yield increase. However, the increase of atmospheric CO_2 side effects will offset any benefits. Public policies for mitigation of climate change will also affect how agriculture is done. It is estimated that agriculture is responsible for roughly 10–12 % of all greenhouse gas emissions. By international accords greenhouse gas must be reduced by 2050 to 50–60 % of current levels, to avoid more serious impacts on the environment (Garnett 2008). The challenge is, therefore, feed the world meeting the environmental agenda (Fuchs et al. 2009).

Before the Industrial Revolution, the atmospheric CO_2 concentration was 280 ppm and today it is 384 ppm (IPCC 2007). It is expected that this carbonic gas concentration will reach between 720 and 1.030 ppm by 2100, and its consequence for agriculture, via Global Warming, will be even more alarming.

According to the perspectives on climate change presented by the Intergovernmental Panel on Climate Change (IPCC 2007) there has been many speculations of what will happen in the coming decades. The most striking aspect is the effects of the climate change on agriculture and its impact on food security, as well as on the economic losses in agribusiness (Assad and Pelegrino 2007). Several models have been developed showing how climate changes may occur and how it may bring adverse effects on agriculture (Lobell et al. 2008; Buntgen et al. 2011).

Some scientists argue that, if greenhouse gas emission continues at present levels, the world temperature would increase between 2.5 and 4.3 °C by 2099 (Christensen et al. 2007). In spite of the dramatic negative effects of Global Warming, in some regions the temperature increase may have a temporary benefit. In high latitude regions crops like corn, wheat, and rice may benefit from warmer climate due to higher CO_2 concentration and rain precipitation (Easterling et al. 2007). These same authors, however, suggest that in regions of low latitude, in the tropics, the increase in temperature will reduce yield. It is believed by some scientist that yield for corn and soybean in the USA maybe reduced by 30–46 % before the end of the present century if the current scenario persists (Kucharik and Serbin 2008).

In general, Global Warming may bring many adverse effects, such as accelerated plant growth, reduced plant phenological cycle, and increasing floral abortion and grain filling among others. Wahid et al. (2007) presents an in-depth review about Global Warming outcome for agriculture. Rain precipitation will be altered in quantity and distribution with bad consequences to plant growth as described in Chap. 6 More problems due to insects and other biotic stresses also expected. In Brazil, agricultural growing zones will have profound changes. For example, the coffee crop grown in many Minas Gerais State regions will move southward.

Arisworth and Ort (2010) discuss several physiological and management alternatives that may be considered to offset the adverse effects of Global Warming on agriculture and help it keep pace with the world demand for food, feed, fiber, and fuel. Some scientists believe that the climate change will be so drastic that currently available technologies may be ineffective. Assad and Pinto (2008), for example, suggested that it is possible that the speed with which climate changes occur makes the conventional methodologies in agriculture become obsolete and ineffective to allow agriculture to cope with the coming challenges.

The historical contribution of plant breeding to agriculture may suggest that those pessimistic forecasts could be avoided. For example, cauliflower, a species adapted to temperate climate has been bred in Brazil for heat tolerance. Up to the 1970s cauliflower was grown in high altitude regions in Rio de Janeiro State. The plant breeder Marcelo Dias introduced heat tolerance from different plant introductions and after a few selection cycles he developed the cultivar Piracicaba Precoce 1. This cultivar allowed this crop to be planted into previously marginal regions for this species. Today, cauliflower is grown in areas with mean temperature much higher than what it is expected to have with Global Warming, showing that plant breeding has a great potential for man to deal with the upcoming challenges.

Carrot is another good example of breeding for heat tolerance in the tropics. Similarly to what happened to cauliflower, in Brazil carrots were grown only in the Brazilian southern states up to the 1970s; but after Embrapa, the Brazilian agency for agricultural research identified carrot populations with heat tolerance; it launched the cultivar Tropical (Costa 1974) that broaden the adaptation region for carrots all over the country (Vieira 1976).

The last example among many other well-known ones is the heat tolerant apple cultivars that also developed in Brazil. Until 1970 all apples sold in Brazil were imported from Argentina and other countries, once the cultivars available in country were not well adapted. Apple cultivars require a minimum of 800 h of hibernation with temperatures under 7.2 °C, a condition just about inexistent in Brazil. Breeding programs targeting clonal selection developed cultivars with a much shorter hibernation requirement. Table 1.5 presents a list of apple clones developed up to 1988, showing cultivars substantially more tolerant to high temperatures (Denardi and Camilo 1998). It was also remarkable that this breeding progress followed by adequate crop management allow this crop in Brazil. Brazil

Table 1.5 Apple cultivars developed in Brazil

Princesa	400–500 h
Primicia	600–650 h
Catarina	>800 h
Baronesa	500–500 h
Condessa	350–400 h
Lisgala	>800 h—Mutation on cv. Gala
Fugi Suprema	>800 h—Mutation on cv. Gala

Source Adapted from Dernard and Camilo (1998)

today is self-sufficient in apple production and also exports this fruit to the European market.

It is important to realize, in the scenario of climate change that plant breeding can and will give an important tool for food security. The development of cultivars adapted to abiotic stresses is the best and most sustainable alternative to support agriculture today and in the future.

1.4 Plant Breeding and the Soil Abiotic Stresses

The soil has been called the *life's placenta*, since it is the source of all plants, which feed the ecosystem (Rao and Cramer 2003). Therefore, the lack or excess of the mineral nutrients in the soil affects the plant development. Soil stresses come about for many different reasons, such as poor soil management, soil genesis, etc.

Many management technologies have been developed to counterpoint stress from soil. For example, in many countries no-till has been widely adopted by farmers. The conventional planting system involves the use of moldboard plow for primary tillage followed by several secondary tillage using disc harrow, and for each 1 ton of corn harvest using this obsolete but still used system it lost 5 tons of top soil due to erosion (Paterniani 1990). With this erosion the soil loses most of its nutrients and the organic matter. Additionally, erosion causes other problems such as sedimentation of soil particles in rivers. In the no-till system the seeds are placed directly into the previous season's crop residue, reducing soil erosion.

One of the most important factors for the great grain yield gains in the 1950s was the use of chemical fertilizers especially in the developing countries. However, plants need six nutrients in larger quantities, called macronutrients and eight in smaller amounts, called micronutrients, the emphasis in this book will be on just two macronutrients: nitrogen and phosphorous, due their shortage and high costs. So, development of cultivars with better nitrogen and phosphorous use efficiency are essential for agriculture. Chapter 4 and 5 address different strategies for cultivar development by those traits.

High soil salinity is also a serious problem in many regions around the world. Salinity in general is caused by sodium salts, in special sodium chlorite, which causes severe problem to most crops. According to the FAO Land and Plant Nutrition Management Service, the more than 6 % of world's land is affected by

Table 1.6 World distribution of soil salinity, in million hectares

Regions	Total area	Saline soils		Sodic soils	
	Mha	Mha	%	Mha	%
Africa	1,899	39	2.0	34	1.8
Asia, the Pacific, and Australia	3,107	195	6.3	249	8.0
Europe	2,011	7	0.3	73	3.6
Latin America	2,039	61	3.0	51	2.5
Near East	1,802	92	5.1	14	0.8
North America	1,924	5	0.2	15	0.8
Total	12,781	397	3.1	434	3.4

Source FAO Land and Plant Nutrition Management Service

salinity or sodicity (Table 1.6), comprising over 400 million hectares. Of the current 230 million ha of irrigated land, 45 million ha are salt-affected.

Plant response to salinity involve modification in several physiological and biochemical processes. The plant's ability of detoxification from free salt ions requires a lot of energy, but it is the best way to grow susceptible crops in marginal due to salinity. Molecular genetics has provided new insights about plant tolerance to salinity. A cell type-specific expression of *AtHKT1;1*, a sodium transporter, has been shown to improve sodium (Na^+) exclusion and salinity tolerance in Arabidopsis (Plett et al. 2010). The authors found that, in the transgenic Arabidopsis plants overexpressing *AtHKT1;1* in the cortex and epidermis, the native *AtHKT1;1* gene responsible for Na^+ retrieval from the transpiration stream, was also unregulated. Extra Na^+ retrieved from the xylem was stored in the outer root cells and was correlated with a significant increase in expression of the vacuolar pyrophosphatases (in Arabidopsis and rice) the activity of which would be necessary to move the additional stored Na^+ into the vacuoles of these cells. This and many other recent studies present important steps in the development of salinity tolerant crops. This topic will be covered on Chap. 7.

When soil is acidic, which is with low pH, aluminum is dissolved in the soil solution and becomes toxic to most plants. Aluminum toxicity limits crop production in as much as half the world's arable land, mostly in developing countries in Africa, Asia, and South America. In Brazil, it comprises 25 % of all agricultural area. Aluminum toxicity affects root growth, reducing the plant ability to exploit the soil water and nutrients. Many biotechnology groups are cloning aluminum-tolerant genes and expect to have tolerant varieties. Breeding for aluminum toxicity tolerance will be discussed in depth in Chap. 8.

1.5 Perspectives

According to a panel of experts by United Nations the world's population will reach 9.1 billion people around 2050. The challenges for agriculture and especially for food production are huge, requiring from plant breeders' high dedication and solid knowledge to develop cultivars fit for those challenges. In the next chapters, different experts will address the most relevant achievements and breeding methodologies applied to cultivar development to produce under abiotic stresses.

References

Ainsworth EA, Ort DR (2010) How do we improve crop production in a warning world? Plant Phisiology 154:526–530

Alexandratos N (2006) World Agriculture: towards 2030/50. interim report. An FAO perspective. Rome: FAO

Assad ED, Pelegrino GO (2007) O clima e a potência ambiental. Agroanalysis 27(9):E3–E5

Assad ED, Pinto HS (2008) Aquecimento global e a nova geografia da produção agrícola no Brasil. São Paulo, Embrapa/Unicamp, p 84

Beddington I (2010) Food security: Contributions from science to a new and greener revolution. Philosophical Transcections Of The Royal Society 365:61–71

Borém A, Almeida GD (2011) Plantas geneticamente modificadas: desafios e oportunidades para regiões tropicais. Suprema, Visconde de Rio Branco

Borém A, Rios SA (2011) Milho biofortificado. Suprema, Visconde de Rio Branco

Boyer JS (1982) Plant productive and environment. Science 218:443–448

Bruinsma J (2009) The resource outlook to 2050: by how much do land. water and crop yields need to increase by 2050? Expert Meeting on How to Feed the World in 2050. Rome. FAO

Buntgen U et al (2011). 2500 Years of European climate variability and human susceptibility. Science Express. 13 Jan, 2011. pp 1–4.

Chris S, Briscoe J (2001) Genetic engineering and water. Science 292: p 2217

Christensen TR, Johansson T, Olsrud M, Strom L, Lindroth A, Mastepanov M, Malmer N, Friborg T, Crill P and Callaghat TV (2007) A catchment-scale carbon and greenhouse gas budget of a subarctic landscape. Phil Trans R Soc A 365:1643–1656

Costa CP (1974) Cenoura Nacional. em germoplasma para as condições de dias curtos nas regiões tropicais e sub-tropicais. Relatório científico do Departamento de Genética. ESALQ/USP. N°8. pp 50–53

Demeke M, Pangrazio G, Maetz M (2008) Country responses to the food security crisis: Nature and preliminary implications of the policies pursued. FAO, Rome

Denardi F, Camilo AP (1998) Estratégia do Melhoramento Genético da Macieira na EPAGRI S. A. Santa Catarina. Anais do II Simpósio sobre atualizações em Genética e Melhoramento de Plantas. UFLA. pp 123–132

Duvick D, Cassman KG (1999) Post-green revolution trends in yield potential of temperature maize in the North-Central United States. Crop Sci 39:1622–1630

Duvick DN, Smith JSC, Cooper M. Long Term selection in a commercial hybrid maize breeding program. In: Janick. I. Plant Breeding Reviews. Part. 2 24:109–152

Easterling WE, Aggarwal PK, Batima P, Brander KM, Erda L, Howden SM, Kirilenko A, Morton J, Soussana JF and Schmidhuber J (2007) Food, fiber and forest products. In: Solomon S, Qin D, Manning M, Chen Z, Marquis M, Averyt KB, Tignor M, Miller HL (eds) Climate Change 2007: The Physical Science Basis. Contribution of Working Group I to the Fourth Assessment Report of the Intergovernmental Panel on Climate Change. Cambridge University Press, Cambridge, UK, and New York, pp 273–313

Evans L (1998) Feeding the Ten Billion: Plants and Population Growth. Cambridge University Press, Cambridge

Evans LT, Fisher RA (1999) Yield potential: Its definition measurement and significance. Crop Sci 39:1544–1551

FAO (2009) How to feed the world in 2050. FAO, Rome

FAO (2010) The state of food insecurity in the world 2010. FAO, Rome
Fuchs D, Kalfagianni J, Artenson M (2009). Retail power. Private standards and sustainability in the Global Food System. In: Clapp J, Fuchs D (eds) Corporate power in global agrifood governance. MIT Press, Cambridge
Garnett T (2008) Cooking up a storm: Food. greenhouse gas emissions and our changing climate. Guildford: food climate research network. Centre for Environmental Strategy. University of Surrey
IPCC (2007) Climate change—the physical sciences basis: summary for policymakers. Intergovernmental Panel on Climate Change, Genebra p 18
Kucharik CJ, Serbin SP (2008) Impacts of recent climate change on Wisconsin corn and soybean yield trends. Environ Res Lett 3:034003 (p 10)
Lobell D, Burke MB, Tebaldi C, Mastrawdrea MD, Falcon WP, Naylor RL (2008) Prioritizing climate chance adaptation needs for food security in 2030. Science 319:607–610
Mitchell R, Kenneth PV, and Gautam S (2008) Managing and enhancing switchgrass as a bioenergy feedstock. Biofuels Bioprod Bioref 2:530–539
Ort D, Long, SP (2003) Converting solar energy into crop production. In: Chrispeels MJ, Sadava. D. E. Plants genes and crop biotechnology. 2. ed. Londres: Jones and Bartlett Publishers International. pp 240-269
Paterniani E (1990) Breeding in the tropics. Critical. Reviews in Plant Science. 9(2):125–154
Peng S, Cassman KG, Virmani SS, Sheehy I, Khush GS (1999) Yield potential trends of tropical rice since the release of IR 8 and the challenge of increasing rice yield potential. Crop Sci 39:1552–1559
Plett A, Safwat G, Gilliham M, Møller IS, Roy SJ, Shirley N, Jacobs A, Johnson A, Tester M (2010) Improved salinity tolerance of rice through cell type-specific expression of AtHKT1. plos One 5:e12571
Quam M (2009) The economics of genetically modified crops. Annual Rev Resour Econ 1:665–694
Ramalho MAP, Silva GS, Dias LAS (2009) Genetic plant improvement and climate changes. Crop Breeding and Applied Biotechnology 9:189–195
Rao IM, Cramer GR (2003) Plant nutrition and crop improvement in adverse soil conditions In: Chrispeels MJ, Sadava DE. Plants genes and crop biotechnology. 2. ed. Londres: Jones and Bartlett Publishers International. pp 270–303
Specht JE, Hume DJ, Kumudini SV (1999) Soybean yield potential. A genetic and physiological perspective. Crop Sci 39:1560–1570
Tollenaar M, Lee EA (2002) Yield potential. yield stability and stress tolerance in maize. Field Crops Res 75:161–169
Tollenaar M, Wu I (1999) Yield improvement in temperate maize is attributable to greater stress tolerance. Crop Sci 39:1597–1604
UK Government (2011). Foresight project on global food and farming futures: trends in food demand and production. p 39
USDA (2009) USDA agricultural projections to 2018. Washington, USDA
Vencovsky R, Ramalho MAP (2006) Contribuições do melhoramento genético no Brasil. In: Paterniani E (Org.). Ciência. agricultura e sociedade. 1.ed. Brasília: EMBRAPA. pp 41–74
Vieira C (1976) Feijão. Viçosa, Impressa UFV p 129
Washid A, Gelani S, Ashraf M, Foolad MR (2007) Heat tolerance in plants: An overview. Environ Exp Bot 61:199–223
Wright J (2010) Feeding nine billion in a low emissions economy—Simple. Though Not Easy. London, A review for the Overseas Development Institute

Chapter 2
Breeding for Stress-Tolerance or Resource-Use Efficiency?

Roberto Fritsche-Neto and Júlio César DoVale

Abstract In plant breeding programs, it is not entirely clear whether the goal is to improve resource-use efficiency or the tolerance to low resource availability. The aim of this chapter was to verify the relationship between breeding for a tolerance to low levels of soil nutrients and for nutrient-use efficiency. The relationship between nutritional efficiency and the tolerance to nutritional stress was estimated by the Spearman ranking correlation between maize genotypes for the traits related to nitrogen, phosphorus and water-use efficiency and phenotypic plasticity indices. The lack of relationship between the traits, in both magnitude and significance, indicates that these traits are controlled by different gene groups. Consequently, simultaneous selection for both nutrient-use efficiency and tolerance to nutritional stress is possible if the mechanisms that confer efficiency and tolerance are not competitive.

Keywords Plant breeding · Abiotic stress

2.1 Introduction

To meet the rising worldwide demand for food, there are some options, such as further exploration of planting in dry seasons, that often leads to lower yields, increased productivity, and the expansion of cultivation areas, an alternative that

R. Fritsche-Neto (✉) · J. C. DoVale
Department of Plant Science, Federal University of Viçosa,
Viçosa, Minas Gerais, Brazil
e-mail: roberto.neto@ufv.br

J. C. DoVale
e-mail: julio.vale@ufv.br

R. Fritsche-Neto and A. Borém (eds.), *Plant Breeding for Abiotic Stress Tolerance*,
DOI: 10.1007/978-3-642-30553-5_2, © Springer-Verlag Berlin Heidelberg 2012

ensures greater food production and is still viable in some emerging countries, including Brazil. Within this context, a significant fraction of cultivation can be performed in marginal areas, which tend to have low fertility (Giaveno et al. 2007).

When water availability and soil nutrients are limiting to plant growth and development, there is a reduction of metabolism, biomass, and the surface areas of various plant organs, thus affecting the productivity (Sultenfuss and Doyle 1999). Considering this fact, several plant breeding programs have emerged, and other lines of research have been directed at improving conditions for abiotic stresses. Thus, two breeding strategies can be considered: tolerance to a low availability of water and nutrients and resource-use efficiency.

2.2 Tolerance and Efficiency as Responses to Abiotic Stress

The literal meaning of the word "stress" is coercion, that is, force in one direction. The term, in Physics, is used to denote the tension produced within a body by the action of an external force. Biologically, stress is considered as a significant deviation from the ideal conditions in which plants are grown, preventing them from expressing their full genetic potential for growth, development, and reproduction (Rehman et al. 2005). According to Larcher (2006), in principle, the change-induced responses at all functional levels of the organism are reversible (elastic deformation) but may become permanent (plastic deformation) (Fig. 2.1).

It has been observed that brief exposure to stress does not exceed the tolerance limit and causes only temporary changes in the plant, whereas prolonged exposure results in permanent changes. Thus, after recovery, the dry matter returns to the original rate (angle of inclination α). However, in the case of chronic stress, the growth rate is reduced at a continuous angle ($\beta < \alpha$), and the loss in productivity is significantly higher.

Abiotic stress may trigger a series of responses in plants that include changes in gene expression and cellular metabolism. The duration, severity, and frequency with which a stress is imposed, and the affected organs and tissues, developmental stage, and genotype also influence plant responses to stress. Consequently, a combination of different conditions can cause different plants responses to the same type of stress (Fig. 2.2).

For the efficient improvement of this process, it is essential to identify the responses that promote or maintain the growth and development of plants during the stress period (Bray et al. 2000).

The tolerance to a particular stress is quite variable but is related to the plant's ability to withstand adverse conditions, survive, and reproduce successfully. Agronomically, Miti et al. (2010) defined tolerance as the reduction in yield under stress conditions compared to the yield under the optimal condition of cultivation.

The use efficiency (UE) of water or nutrients is defined as the ratio between the yield per unit of resource available to the plant (Moll et al. 1982). In studies with plants in early stages, the yield is usually replaced by the mass of shoot dry weight

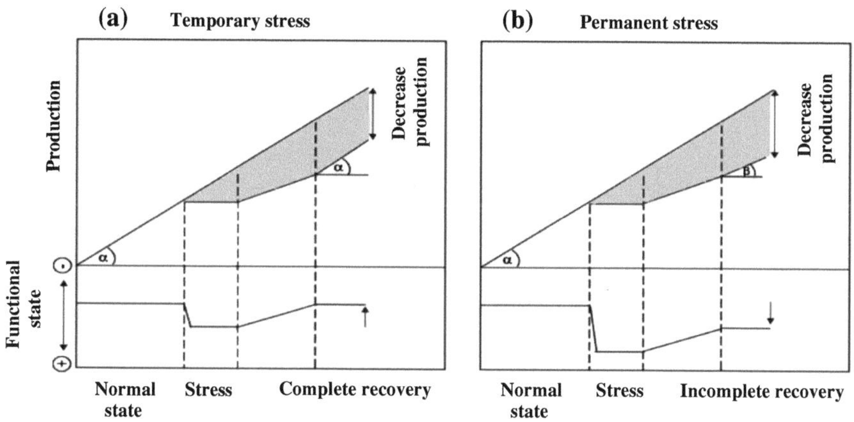

Fig. 2.1 Effect of environmental stress on productivity. **a** Temporary stress and **b** Permanent stress. *Source* Adapted from Larcher (2006)

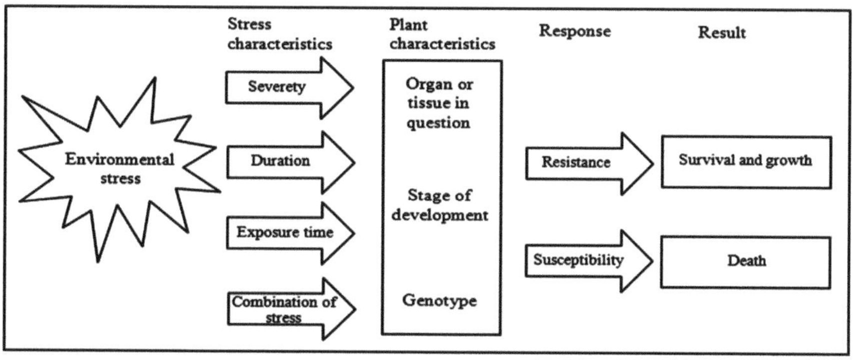

Fig. 2.2 Plant responses in accordance with the characteristics of environmental stress. *Source* Bray et al. (2000)

to estimate the UE (Maia et al. 2011; Fritsche-Neto 2011). Thus, particular genotypes will be considered as efficient when producing well with the minimum availability of the resource (Good et al. 2004).

The UE consists of the acquisition efficiency (AE) and utilization efficiency (UtE). The AE is defined as the ability of a given genotype to absorb certain nutrients or soil water, whereas the UtE refers to the ability of this genotype to produce biomass or grain using the resource that has been absorbed. For Hirel et al. (2007), genotypes more efficient in the use these resources can be achieved by increasing the AE and UtE.

In cases of tolerance and efficiency, plants use physiological mechanisms, and sometimes, anatomic mechanisms to avoid the effect of stress and rapidly recover (Zheng et al. 2000). As a result, three main strategies have been recognized that

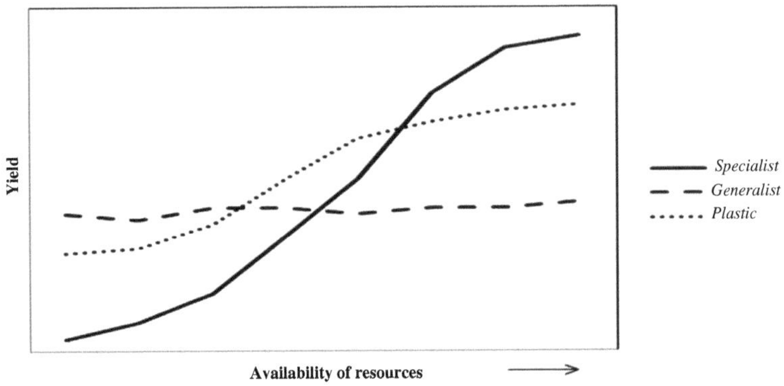

Fig. 2.3 Classification and behavior of plants based on their response to heterogeneity in resource availability

plants use to cope with stress: (*i*) *specialization*, the genotype is adapted to the specific environment; (*ii*) *generalization*, the genotype has moderate suitability in most environments; and (*iii*) *phenotypic plasticity*, signals from the environment interact with the genotype and stimulate the production of alternative phenotypes (Fig. 2.3).

The farmer wants cultivars that produce a satisfactory yield when subjected to stress conditions but that have a high productivity under ideal growing conditions. However, considering the concepts stated above, from a physiological point of view, the simultaneous improvement of abiotic stress tolerance and resource-use efficiency is contradictory. Furthermore, there is some confusion between these terms in many reports published and also in the objectives of research.

According to Bradshaw (2006), phenotypic plasticity is high when compared to the yield stability. Thus, low plasticity (or high stability) is not always a desirable characteristic because tolerant genotypes generally have moderate productivity, even under ideal growing conditions (Cruz et al. 2004), which is important in marginal areas and under cultivation with permanent stress. Thus, these genotype "generalists" exhibit superior productivity in environments with low resource availability. However, when the environmental conditions are not limiting, there is no significant increase in productivity, which is a disadvantage.

Given this fact, most of the improvement programs are aimed at increasing the UE or in obtaining genotypes with high phenotypic plasticity. Thus, when the plants are subjected to limiting conditions, they would use fewer resources to produce satisfactory results but show high yields when the conditions are ideal.

The genetic control of both stress tolerance and resource-use efficiency is quantitative and involves many loci distributed in different regions of the genome in cultivated species (Wu et al. 2011). Quantifying and understanding the genetic relationship between these two strategies for improvement in plant selection is the key in the implementation of programs to improve and increase the effectiveness of the approach.

Table 2.1 Spearman correlation estimates between the plasticity index of shoot dry weight mass (iSDW), length lateral roots (iL_{LAT}), axial roots (iL_{AXI}) and ratio shoot/root (iRSR) to the acquisition efficiency (AE) and utilization efficiency (UtE) at high (HN) and low nitrogen (LN) and high (AP) and low phosphorus (LP) in 15 tropical maize inbred lines

Trait	AE_{LN}	AE_{HN}	UtE_{LN}	UtE_{HN}	AE_{LP}	AE_{HP}	UtE_{LP}	UtE_{HP}
iSDW	−0.11[a]	0.18[a]	0.28[a]	0.49[a]	0.22[a]	0.15[a]	0.04[a]	−0.23[a]
iL_{LAT}	0.28[a]	0.41[a]	−0.25[a]	−0.27[a]	0.28[a]	−0.20[a]	0.02[a]	0.08[a]
iL_{AXI}	0.50[a]	0.11[a]	−0.14[a]	0.07[a]	0.42[a]	0.26[a]	0.12[a]	0.11[a]
iRSR	−0.23[a]	0.05[a]	0.37[a]	−0.35[a]	−0.28[a]	−0.40[a]	−0.06[a]	0.00[a]

[a] Not significant at $p < 0.05$ % by the t test

Source Adapted from Maia (2011)

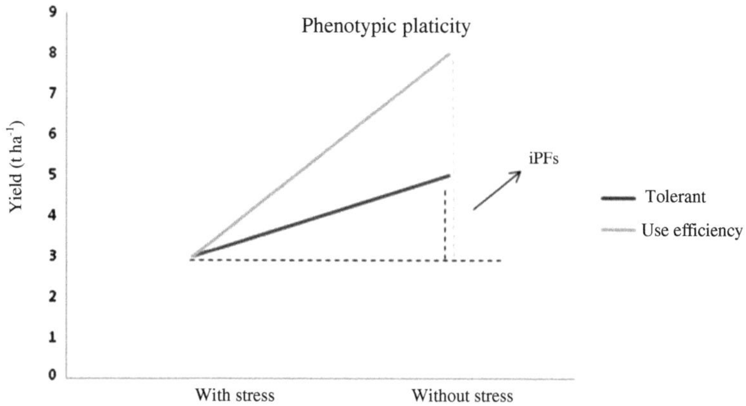

Fig. 2.4 Phenotypic plasticity indexes (iPFs) of two hypothetical genotypes, which demonstrate contrasting behaviors in relation to stress (with stress and without stress)

2.3 Results Observed

Seeking to quantify the possible relationship between improvements in stress tolerance and nutrient efficiency in corn, Maia et al. (2011) evaluated strains in two experiments conducted in a greenhouse under conditions of contrasting nitrogen (N) and phosphorus (P). The relationship between the nutritional efficiency and tolerance to the respective nutrients was estimated using the *Spearman* correlation between the rank of the lineages selected for traits related to UE of N and P or the rate of phenotypic plasticity (Valladares et al. 2000) (Table 2.1).

There is a lack of correlation among the traits, both for significance and magnitude, indicating that they are controlled by different gene groups. Therefore, mechanisms that provide an efficient use of resources and stress tolerance and are not competitive will allow the simultaneous selection for both traits. However, one must consider that most of the observed correlations were negative. With the increase in effective size, the coefficients would have a greater magnitude, and its effects, most likely, would be significant.

Table 2.2 Means of the plasticity index and water-use efficiency (WUE) for five strains of tropical maize

Maize lines	Plasticity index	WUE
13	0.60	1.465
9	0.59	1.535
12	0.57	1.43
3	0.46	1.405
7	0.44	1.335

Source Adapted from Coutinho (2011)

In another study, Wu et al. (2011) observed differences in the performance of 189 maize lines under contrasting conditions of N on the agronomic UE of this nutrient and the index of tolerance to a low soil N. The authors concluded that there is genetic variability for both the N-use efficiency and tolerance to this stress.

The physiological point of view predicts that the genotypes with higher resource-use efficiencies are not the most tolerant to the shortage of that resource. Thus, considering the rate of phenotypic plasticity, genotypes are classified as tolerant when they have the lowest values of this index, that is, the lower range of productivity between environments with and without stress. Genotypes with higher values for the index of plasticity, in turn, are considered efficient in resource usage or more efficiently adaptive to changes in growing conditions. For these genotypes, there are considerable increases in productivity with small increases in resource availability (Fig. 2.4).

A study that supports this evidence was conducted by Coutinho (2011) in the selection among 14 tropical maize inbred lines that contrasted with the parents for water-use efficiency (WUE). According to the author, the lines that showed the lowest rates were those with less WUE plasticity, whereas the lines that showed higher rates were those with more WUE plasticity (Table 2.2).

It is believed that, in most cases, genotypes that are efficient under nutritional stress and water stress should be sought, i.e., those that respond positively to the addition of features that enhance productivity. However, other factors, such as restrictive cultivation, temperature, content of aluminum, and salinity, should also be considered when selecting more tolerant genotypes.

Finally, we emphasize the need of additional studies for better understanding of the relationship between resource-use efficiency and stress tolerance to increase the efficiency of breeding programs for plants growing under conditions of abiotic stress.

References

Bradshaw AD (2006) Unravelling phenotypic plasticity—why should we bother? N Phytol 170:644–648

Bray EA, Bailey-Serres J, Weretilnyk E (2000) Responses to abiotic stresses. In: Buchanan BB, Gruissem W, Jones RL (eds) Biochemistry and molecular biology of plants. American Society of Plant Physiologists, Rockville, p 1367

Coutinho PH (2011) Herança e relação dos caracteres fisiológicos e de raízes com a eficiência no uso da água em milho. UFV, Viçosa, p 48

Cruz CD, Regazzi AJ, Carneiro PCS (2004) Modelos biométricos aplicados ao melhoramento genético, 3rd edn, vol 1. Editora UFV, Viçosa, p 480

Fritsche-Neto R (2011) Seleção genômica ampla e novos métodos de melhoramento do milho. UFV, Viçosa, p 28

Giaveno CD, Ribeiro RV, Souza GM, Oliveira RF (2007) Screening of tropical maize for salt stress tolerance. Crop Breed Appl Biotechnol 7:304–313

Good AG, Shrawat AK, Muench DG (2004) Can less yield more? Is reducing nutrient input into the environment compatible with maintaining crop production? Trends Plant Sci 9:597–605

Hirel B, Le Gouis J, Ney B, Gallais A (2007) The challenge of improving nitrogen use efficiency in crop plants: towards a more central role for genetic variability and quantitative genetics within integrated approaches. J Exp Bot 58:2369–2387

Larcher W (2006) Ecofisiologia vegetal. Rima, São Carlos, p 550

Maia CB, DoVale JC, Fritsche-Neto R, Cavatte PC, Miranda GV (2011) The difference between breeding for nutrient use efficiency and for nutrient stress tolerance. Crop Breed Appl Biotechnol 11:270–275

Miti F, Tongoona P, Derera J (2010) S_1 selection of local maize landraces for low soil nitrogen tolerance in Zambia. Afr J Plant Sci 4:67–81

Moll RH, Kamprath EJ, Jackson WA (1982) Analysis and interpretation of factors which contribute to efficiency of nitrogen utilization. Agron J 74:562–564

Rehman MS, Haris PJC, Ashraf M (2005) Stress environments and their impact on crop production. In: Ashraf M, Harris PJC (eds) Abiotic stress—plant resistance through breeding and molecular approaches. Food Products Press, New York, p 725

Sultenfuss JH, Doyle WJ (1999) Phosphorus for agriculture. Better Crops Plant Food 83:1–40

Valladares F, Wright JS, Lasso E, Kitajima K, Pearcy RW (2000) Plastic phenotypic response to light of 16 congeneric shrubs from a panamanian rainforest. Ecology 8:1925–1936

Wu Y, Liu W, Li X, Li M, Zhang D, Hao Z, Weng J, Xu Y, Bai L, Zhang S, Xie C (2011) Low-nitrogen stress tolerance and nitrogen agronomic efficiency among maize inbreds: comparison of multiple indices and evaluation of genetic variation. Euphytica 178 (online first)

Zheng HG, Babu RC, Pathan MS, Ali ML, Huang N, Courtois B, Nguyen HT (2000) Quantitative trait loci for root penetration ability and root thickness in rice: comparison of genetic backgrounds. Genome 43(1):53–61

Chapter 3
The Physiology of Abiotic Stresses

Paulo C. Cavatte, Samuel C. V. Martins, Leandro E. Morais, Paulo E. M. Silva and Fábio M. DaMatta

Abstract Plants are often exposed to several adverse environmental conditions that potentially generate stress and thus negatively affect their growth and productivity. Understanding the physiological responses of crops to stress conditions is essential to minimizing the deleterious impacts of stress and maximizing productivity. Therefore, there is urgent need for more scientific research to increase our understanding of the physiological behavior of crops in response not only to a specific type of stress but also to multiple interacting stressors, such as water, and thermal stresses. The proper assessment of this information may result in important tools for monitoring the most promising genetic material in plant breeding programs. In this chapter, the plant strategies associated with satisfactory growth and yield under abiotic stress conditions are discussed, with emphasis in tropical environments. In addition, the state of the art on the physiology of the major abiotic stresses (drought, salinity, heat, nitrogen and phosphorus deficiencies and aluminum toxicity) and possible strategies to develop cultivars with

P. C. Cavatte · S. C. V. Martins · L. E. Morais · P. E. M. Silva · F. M. DaMatta (✉)
Departamento de Biologia Vegetal, Universidade Federal de Viçosa,
Avenida Peter Henry Rolfs s/n, Viçosa-MG, 36570-000, Brazil
e-mail: fdamatta@ufv.br

P. C. Cavatte
e-mail: cavattepc@hotmail.com

S. C. V. Martins
e-mail: samuelmartins25@yahoo.com.br

L. E. Morais
e-mail: leandroufjf@yahoo.com.br

PauloE. M. Silva
e-mail: paulomenezesbio@gmail.com

R. Fritsche-Neto and A. Borém (eds.), *Plant Breeding for Abiotic Stress Tolerance*,
DOI: 10.1007/978-3-642-30553-5_3,

satisfactory productivity in stressful environments using a physiological approach are summarized.

Keywords Aluminum · Breeding · Drought · Heat · Nitrogen · Phosphorous · Salinity · Stress · Yield

3.1 Introduction

Because of the present scenario of global climate changes and considering that major advances in agriculture were designed for environments favorable to the "Green Revolution", crop performance under adverse conditions in marginal environments, which has often been overlooked, is currently the subject of constant debate. Plants are often exposed to several harsh environmental conditions that potentially generate stresses. Stresses trigger a wide range of plant responses, from altered gene expression and cellular metabolism to changes in growth rates and crop yields. The duration, severity, and rate at which a stress is imposed all influence how a plant responds. A response may be triggered directly by a stress, such as water deficit, or may result from a stress-induced injury, such as loss of membrane integrity (Bray et al. 2000).

The indubitable importance of abiotic stresses to agriculture can be easily perceived taking into consideration that altogether abiotic factors provide the major limitations to crop yield worldwide, and can reduce average productivity of annual crops by 51–82 %, depending on the crop (Bray et al. 2000). Understanding the physiological responses of crops to stress conditions is, thus, essential to minimizing the deleterious impacts of stress and maximizing productivity. Therefore, there is urgent need for more scientific research to increase our understanding of the physiological behavior of crops in response not only to a specific type of stress but also to multiple interacting stressors, such as water and thermal stresses. The proper assessment of this information may result in important tools for monitoring the most promising genetic material in plant breeding programs.

In this chapter, the plant strategies associated with satisfactory growth and yield under abiotic stress conditions are discussed, with emphasis in tropical environments. In addition, the state of the art on the physiology of the major abiotic stresses [(drought, salinity, heat, nitrogen (N) and phosphorus (P) deficiencies, and aluminum toxicity)] and possible strategies to develop cultivars with satisfactory productivity in stressful environments using a physiological approach are summarized herein.

3.2 Water Deficit

In recent years, losses related to drought have been the main challenge for grain production, and it is the environmental factor that most limits crop productivity worldwide, especially in semi-arid regions. Drought events have become stronger in recent decades and are most likely associated with climate changes due to global warming. According to climate projections, this situation tends to be exacerbated, and thus, the availability of water for irrigation is expected to be reduced (IPCC 2007). The demand for food, however, is growing significantly, such that more water will be necessary for agricultural activities in the future. In this context, the use of water (consumption) and efficiency of its use (water use efficiency, WUE; i.e., amount of water consumed to produce a given amount of biomass) are key parameters in areas affected by water deficit. Indeed, one of the major goals of modern agriculture is to achieve increased crop productivity using less water ("more crop per drop").

For purposes of this chapter, drought must be defined. In a broad sense, drought is a multidimensional phenomenon, including water deficits in not only the soil but also the atmosphere, which is primarily determined by relative humidity and air temperature. In this chapter, drought will be considered as a meteorological phenomenon, described as a sufficiently long period of imbalance between precipitation and evapotranspiration, that is capable of promoting the depletion of soil water and causing water stress in plants, thereby reducing their growth and/or yield.

3.2.1 Plants Strategies Against Water Deficit

The productivity of plants with a water deficit depends on climate and soil conditions, which affect the quantity of water available and how the water is used by the plant. The responses to limited water availability are diverse with respect to the plant species involved and/or the severity and duration of the water deficit. Basically, there are three strategies by which plants can grow and develop properly in environments with water restrictions: (i) drought escape, which can be observed in short cycle crops and which allows the plants to complete their productive cycle before the water deficit becomes severe; (ii) drought avoidance, which, for example, reduces transpiration or increases the absorption of water; and (iii) drought tolerance, which involves some protoplasmic tolerance. Each of these strategies is associated with costs and benefits, which vary according to the species, the environment, the technological resources of the farmer, and the goals of the plant breeding program (Tardieu 2005).

The adaptation of plants to water deficit is a very complex process, involving morphological, physiological, biochemical, and molecular alterations (Passioura 1997). As reasoned by DaMatta (2003), in the short-term soil drought, yield reduction could be related to reduced stomatal conductance (physiological variable linked to the degree of stomata opening) and concomitantly lower photosynthetic rates, but in the long term, a smaller leaf area (due to decreased leaf size and production, and higher rates of leaf senescence) and an altered assimilate partitioning between plant structures and organs could be more directly responsible for decreased crop yield. Additional reductions in stomatal conductance are to be expected as internal water deficits develop and, as a consequence, stomatal limitations to photosynthesis should be exacerbated with the progression of drought, thus further contributing to limit crop yield. Under moderate water deficit conditions, decreases in leaf area are not necessarily accompanied by decreases in photosynthetic rates on a leaf area basis, but if drought progresses further, strong decreases in photosynthetic rates per unit leaf area are often observed (DaMatta 2003). Another problem associated with reduced stomatal conductance (and thus reduced transpiration rate) is increased leaf temperature, which has possible consequences on higher maintenance respiration and photorespiration rates, which further contributes to decrease crop productivity.

In addition to factors leading to a lower transpiration rate, the adaptation to water deficit involves a greater capacity for water absorption by the plant. This process can be accomplished with genotypes that have more robust root systems (deep and branched), which can better exploit deeper soil water reserves (Pinheiro et al. 2005). Indeed, a hallmark of plants subjected to water stress is an increase in the biomass allocated to the root system.

3.2.2 Physiological Strategies for Productivity Increases Under Water Deficit Conditions

Under water stress conditions, the yield (Y) of a crop can be described by the equation (Passioura 1977): $Y = E \times WUE \times HI$, where $E =$ transpired water, $WUE =$ water use efficiency, and $HI =$ harvest index (i.e., relationship between the biomass of the organ of commercial interest and the total plant biomass). Large increases in productivity through breeding programs occurred by increasing the harvest index. For most annual species grown, however, HI is close to its maximum value, and additional increases in crop yields must necessarily be accomplished through increased biomass accumulation. Therefore, the challenge is to increase the capacity of the plant to produce a larger amount of dry matter per area unit. In the case of breeding programs for drought conditions, this increased biomass production must be combined with lower water consumption or increased WUE (Tambussi et al. 2007).

WUE can be estimated easily and quickly using a small amount of dry biomass, which release $^{12}CO_2$ and $^{13}CO_2$ while undergoing combustion in a mass spectrometer. The resulting data can be used to estimate the carbon isotopic discrimination ($\Delta^{13}C$), and the WUE can then be integrated overtime (Farquhar et al. 1989). The high negative correlation between $\Delta^{13}C$ and WUE has been experimentally demonstrated in a large number of species (Condon et al. 2002; Monneveux et al. 2007). Thus, $\Delta^{13}C$ has been a widely used parameter in breeding programs to select genotypes with higher productivity under drought conditions (Ehleringer et al. 1993).

Reducing the stomatal conductance, increasing the photosynthetic capacity or even combining both factors may lead to a higher WUE. A reduction in stomatal conductance is not favorable in breeding programs aimed at increasing productivity because it entails less CO_2 influx and lower photosynthetic rates and, therefore, less biomass accumulation. Thus, the great challenge is to increase the photosynthetic capacity under conditions of low stomatal conductance (Tambussi et al. 2007). Several strategies have been proposed to increase the photosynthetic capacity of cultivars. One of these strategies involves CO_2 concentrator mechanisms, such as the one found in species that utilize C_4 metabolism. Another strategy would be to increase the CO_2 specificity of the enzyme ribulose-1,5-bisphosphate carboxylase/oxygenase (Rubisco), thus reducing losses related to photorespiration in C_3 plants (Parry et al. 2005). The increase in the mesophyll conductance is directly associated with increases in photosynthetic rates without the need for increased stomatal conductance. Finally, another strategy involves increasing the specific leaf mass because such an increase represents a greater amount of photosynthetic apparatus per leaf unit area.

At a cellular level, osmotic adjustment has been considered as the most striking acclimative response of plants subjected to water deficit (Blum 2005), which is characterized by a net solute increase in cells. These solutes, named compatible osmolytes, are organic compounds (e.g., proline, sugar alcohols, sorbitol, mannitol) that do not interfere with cell metabolism. The osmotic adjustment has two main functions: (i) maintaining cell turgor in the presence of negative water potential, which allows the process of cell elongation to be maintained in addition to allowing relatively larger stomatal aperture and (ii) providing a greater capacity for water uptake by the roots, which would also contribute in the maintenance of higher stomatal conductance. Concomitantly, osmotic adjustment leads to the maximization of the photosynthetic process, resulting in relatively greater biomass accumulation. Thus, genotypes with greater potential for osmotic adjustment tend to be more tolerant and, therefore, more productive under drought conditions (Ludlow et al. 1990; Morgan et al. 1992; Moinuddin and Khanna-Chopra 2004).

Additional factors, such as excessive temperature and high irradiance, are common under drought conditions. These factors explain why drought is considered to be a multidimensional stressor (DaMatta 2003). With a reduction in photosynthesis, the use of incident solar radiation decreases, leading to the accumulation of reducing power (NADPH) and a reduction in the pool of the final acceptor ($NADP^+$) of the electron transport chain. This excess reducing power may lead to an over-reduction of the electron transport chain. In this process, electrons may escape and, in turn, react with molecular oxygen to form the so-called reactive oxygen species (ROS; e.g., superoxide anions, hydrogen peroxide, and hydroxyl radicals), which can damage plant metabolism through processes such as lipid peroxidation and protein and nucleic acid oxidation. These oxidative damages can be observed in leaves as chlorotic areas, which in many cases progress to necrosis, leading to leaf abscission. Genotypes with higher antioxidant activity, both enzymatic (increased activity of enzymes, such as superoxide dismutase, catalase, ascorbate peroxidase, and glutathione reductase) and non-enzymatic (e.g., glutathione and ascorbate), tend to be more tolerant to drought, showing less oxidative damage compared with genotypes with less robust antioxidant systems. This high antioxidant capacity confers increased cellular protection, especially for the photosynthetic machinery, and maintenance of leaf area, which consequently results in increased production, even under severe water deficits (Lima et al. 2002; Posch and Bennett 2009).

3.2.3 *Difficulties and Advances in the Development of Drought-Tolerant Cultivars*

The progress performed in the last 100 years, in relation to obtaining drought-tolerant cultivars, has been very weak in most breeding programs, regardless of the species considered, in contrast to the improvements made with respect to other biotic and abiotic stresses (Tardieu 2005). In addition, from a physiological point of view, many experiments have been conducted in greenhouses or growth chambers with plants in small pots, a condition that causes the plants to be exposed to drastic and swiftly imposed drought. Thus, the responses obtained do not entirely reflect the field situation, and therefore, the comparative results should be treated with caution. Moreover, in many cases, the genotypic comparative analyses have been performed with varying hydration degrees and, hence, are not strictly comparable. Consequently, much of the drought tolerance information generated by physiologists has limited utility for the crop breeder because such information cannot be converted to adequate indices to predict the behavior of genotypes with varying levels of tolerance to water stress. The dissociation between basic information and the requirements of applied (or applicable) information is an additional bottleneck for advancements leading to the acquisition of more drought-tolerant materials.

In brief, drought tolerance is the result of several traits that are expressed differently and concomitantly, depending on a variety of factors, including the severity and rate of water deficit imposition, the age and nutritional status of the plant, the soil type and depth, the fruit load status and the atmospheric evaporative demand. Therefore, the adoption of only one strategy for adaptation to drought is certainly inadequate for any type of environment (Sambatti and Caylor 2007). Obviously, numerous genes affect the aforementioned conditions, and therefore, the expected progress in increased drought tolerance through genetic engineering is disappointing. Moreover, only a small proportion of genes involved in plant responses to water deficit are known. Simulating the interaction between genes in complex systems is virtually impossible: for example, a study of 15 genes with three alleles each would generate 14 million genotypes, far too many to be analyzed in detail, particularly without the risk of misinterpretation (Tardieu 2005).

In any event, there is great need to conduct multidisciplinary research involving molecular, physiological, and classic genetic improvement to enhance crop performance in drought-prone regions. Recent advances in the transcriptome, proteome, and metabolome have allowed an integrated view of the expression of thousands of genes and their products involved in drought tolerance (Tuberosa and Salvi 2006). In the long term, these improvements will provide a better understanding of the polygenic nature of drought tolerance. The adequate selection of indices and parameters for the identification of genetic material with increased drought tolerance has resulted in satisfactory increases in the yield of some crop species. Improved WUE, more robust root systems, osmotic adjustments, and increased activity of the antioxidant system are some of the features that should be focused on in breeding programs aiming to increase production under water deficit conditions. Recently, Muuns et al. (2010) pointed out that changes in spectral reflectance and fluorescence related to altered light harvesting and biochemistry caused by oxidative stress may be easily monitored (Munns et al. 2010); a combination of spectral reflectance, fluorescence, and thermal sensing can, thus, provide a powerful tool for large population screening to identify genotypes with improved tolerance to drought stress (Chaerle et al. 2009).

3.3 Salinity

More than 800 million hectares worldwide are estimated to be affected by high soil salinity (FAO 2005). In addition, intensive agriculture and inadequate management practices are causing substantial salinization of agricultural areas. A large proportion of the cultivated regions have become saline due to increased irrigation. When irrigation water contains a high concentration of solutes and there is no possibility of unloading the accumulated salts into a drainage system, these salts can quickly reach levels that are harmful to crops. Thus, salinity is a major threat to the sustainable irrigation necessary to meet the increasing human population demands for food. The importance of breeding programs aiming to increase tolerance to drought and salt stress is therefore inherent.

3.3.1 Salinity Tolerance

Plants are commonly divided into two distinct groups according to the degree of salinity tolerance. Halophyte plants are native to saline soils and complete their life cycle in saline environments, and thus, they are more tolerant to high salt concentrations in the soil. Conversely, glycophytes, or non-halophyte plants, are unable to tolerate the same concentration of salt as halophytes. There is substantial genetic variation regarding salt tolerance both in halophyte and glycophyte plants.

The accumulation of salts in the soil causes a significant reduction in water potential, affecting the water balance of plants and requiring the development of more negative water potentials to maintain the water potential gradient, which allows the movement of soil water into the plant (Munns and Tester 2008). This osmotic effect of dissolved solutes is similar to soil water deficit, and the initial plant responses to excessive levels of salt are the same as described above for water deficit.

In addition to the osmotic effect, salinity leads to an ionic imbalance caused by a high sodium (Na^+) concentration. Metabolic disturbances generated by Na^+ accumulation in cells are partly due to competition with potassium (K^+) for the active sites of enzymes (Blumwald et al. 2000) and ribosomes (Tester and Davenport 2003). K^+ is the activator of over 50 plant metabolic enzymes and cannot be replaced by Na^+ in this function. Thus, high concentrations of Na^+ or a high Na^+/K^+ ratio will affect several critical metabolic processes. High concentrations of specific ions, especially Na^+ and Cl^-, can cause a nutritional imbalance in plants, which frequently involve the deficiency of ions, such as K^+, Ca^{2+}, Mg^{2+}, NO_3^-, and PO_4^- (Läunchli and Epstein 1990).

The plant response to salinity develops according to the limiting factor to growth, and this effect involves an osmotic or ionic imbalance caused by Na^+ excess. Typically, the first phase of salt stress is represented by the osmotic effect that occurs rapidly from the moment that the salt concentration around the roots reaches the tolerance threshold of the species (approximately 40 mM NaCl for most species). In cereals, the first phase of salt stress is marked by a reduction in the number of tillers, whereas in dicotyledons, the main effect is reductions in leaf size and branch number (Munns and Tester 2008). The second stress phase occurs more slowly and involves the per se toxicity of Na^+ ions inside the cell, which is reflected by the accumulation of salt in older leaves, which usually die.

The mechanisms of salinity tolerance are connected to several processes: the recovery of ionic equilibrium and cellular homeostasis, the osmotic adjustment to maintain water absorption by the plant, and the prevention or mitigation of oxidative stress that occurs in the same way as in water deficit. Osmotic adjustment may involve the accumulation of inorganic ions (Na^+ and Cl^-), as typically observed in halophytes, or via the synthesis of compatible osmolytes, a response usually observed in glycophytes (Ashraf 2004). Another type of adaptation to salt stress involves the ability to remove excess salt by specialized salt glands; ions are transported to these glands, where the salt is crystallized and rendered no longer harmful to the plants (Thomson et al. 1988).

For the recovery of cell homeostasis as a function of increased Na^+ levels, the regulation of several carriers involved in the maintenance of appropriate Na^+ levels in the cytosol and the involvement of the salt overly sensitive (SOS) pathway play important roles. The SOS signaling pathway was discovered in *Arabidopsis* and comprises three key components: SOS3, a sensor of Ca^{2+}; SOS2, a serine/threonine protein kinase; and SOS1, a Na^+/H^+ antiporter (exchanger) present in the plasma membrane (Türkan and Demiral 2009). The presence of Na^+ in the apoplast is perceived by an as yet unknown sensor that alters the levels of cytoplasmic Ca^{2+}, activating SOS3 and forming the SOS2/SOS3complex. This complex is responsible for activating SOS1 and the Na^+/H^+ antiporter NHX1 in the tonoplast membrane. These transporters enable Na^+ sequestration in the vacuole and its subsequent extrusion from the cell, reducing the amount of Na^+ in the cytosol. The SOS2/SOS3 complex is also involved in the regulation of the gene expression of the aforementioned transporters and of the activity of vacuolar H^+-ATPase and H^+-PPase, which are responsible for providing H^+ ions for NHX1 activity (Chinnusamy et al. 2005a). Another primary mechanism of salt tolerance, conserved in both monocotyledons and dicotyledons, is performed by HKT transporters, which are responsible for maintaining a high K^+/Na^+ ratio in the leaves. Indeed, a high K^+/Na^+ ratio can be a good marker for salt tolerance (Hauser and Horie 2010). The transporter AtHKT1;1 is responsible for Na^+ removal via the xylem vessels in parallel to the stimulatory effect of K^+ loading, thus controlling the concentration of these two elements to maintain them in a favorable ratio.

The mechanisms mentioned above require an efficient supply of energy in the form of ATP, reducing power and carbon skeletons, either to maintain ATPases or to synthesize compatible osmolytes. This high energy demand has highlighted the role of mitochondrial respiration as an integrator of important mechanisms involved in salt tolerance (Jacoby et al. 2011). Genotypes tolerant to salt stress have been shown to have a more robust mitochondrial antioxidant system, which would be responsible for maintaining the integrity of the energy efficiency required for the acclimation to high salinity conditions. In addition, the ROSs formed in the mitochondria have a signaling role, which may be responsible for orchestrating antioxidant responses in other plant parts. The mitochondrion itself also plays a role in the osmotic adjustment because proline catabolism occurs within mitochondria, with proline abundance affecting the regulation of enzymes involved in mitochondrial catabolism (Jacoby et al. 2011).

The extent of the mechanisms used by salt-tolerant species may differ even intraspecifically. In maize, for example, vacuolar Na^+ compartmentalization occurs in some but not all strains, and conversely, a higher K^+/Na^+ ratio and lower Na^+ content in the tissues are apparently essential characteristics for the tolerance of maize to salinity. In some species, such as *Brassica napus*, however, the inverse situation occurs because the most tolerant genotypes have higher Na^+ concentrations in the aboveground parts but in parallel with a higher accumulation of proline and K^+ (Moller and Tester 2007).

To elucidate the mechanisms associated with salt tolerance, several transgenic approaches have already been performed, although there are few studies in which comparisons are made properly or under field conditions typically experienced by crops (Flowers 2004; Ashraf and Akram 2009). Another complicating factor is the evaluation of the importance of each mechanism involved in salt tolerance. Discussing several examples of salinity tolerance, Ashraf (2004) concluded that the identification of the physiological and biochemical indicators for each species individually is more reliable than the use of generic selection markers due to the variability of the responses at the inter and intraspecific levels. Likewise, caution should be applied when extrapolating results found in model plants, such as *Arabidopsis*, for use in other species because the tolerance strategies are different among groups. Several *Arabidopsis* ecotypes are known to be more salt tolerant via the accumulation of Na^+, whereas most cereals show the opposite response, with an increased tolerance via decreased Na^+ levels in the tissues (Moller and Tester 2007).

The polygenic nature of salt tolerance has resulted in difficulties for the development of improved varieties through conventional genetic improvement, although classical plant breeding has had greater success in the production of new varieties more tolerant to salinity than transgenic breeding strategies (Ashraf and Akram 2009). Therefore, the most effective strategy for increasing salinity tolerance remains the use of the genetic variability within crops. The domestication of halophytes also plays an important role in the use of saline areas, and the use of some agronomic practices, such as increasing the levels of Ca^{2+} in the soil, have been reported to soften the salt effects in some crops (Cimato et al. 2010).

3.4 Heat Stress

The threshold temperature (upper and/or lower), which varies both inter and intraspecifically, refers to the daily average temperature that causes a reduction in growth. The temperature increase above the upper limit for a period of time sufficient to cause irreversible damage to plant growth and development is defined as *heat stress*, which is a complex function of the intensity, duration, and rate of temperature increase (Peet and Willits 1998).

Plants are subjected to heat stress under the following conditions: (i) the air temperature is high, and the plants receive energy by the transfer of sensible heat; (ii) solar radiation incident on the ground raises the temperature above the air temperature; and (iii) solar radiation-induced heating and the inability to dissipate heat can rapidly warm the leaves (up to 15 °C above the air temperature or more), particularly in leaves that have a low transpiration rate and are frequently subjected to high temperatures (Singsaas et al. 1999).

Heat stress can affect crop yield at any time from sowing to grain maturity, but it is the time around flowering, when the number of grains per land area is established, and during the grain-filling stage, when the average grain weight is determined, that high temperatures have the most impact on the final harvestable crop, as found in cereals (Morison and Lawlor 1999; Barnabás et al. 2008). Thus, understanding how environmental factors signal phenological processes such as flowering will be extremely relevant for future food production, since a large part of food comes as grains/seeds. In this context, any change in flowering time could affect not only seed production but also food composition (DaMatta et al. 2010).

3.4.1 Plant Response and Morpho-Physiological Alterations to High Temperatures

Thermal stress increases the kinetic energy and movement of the molecules composing biological membranes, leading to the relaxation of chemical bonds, and providing fluidity to the lipid bilayer. The denaturation of proteins and/or the increase of unsaturated fatty acids also contribute to this process. These changes increase membrane permeability, which is demonstrated by an increased loss of electrolytes. This behavior indicates a decreased thermostability of the cell membrane, which is used as an indirect measurement of heat stress tolerance in several species, such as cotton (Ashraf et al. 1994), sorghum (Marcum 1998), beans (Ismail and Hall 1999), and barley (Wahid and Shabbir 2005).

The membrane properties can be adjusted to increase the temperature range tolerated by plants, such as through changes in fatty acid composition. These properties may be linked with the tolerance of photosynthesis to high temperature stress. In *Arabidopsis* plants grown at high temperatures, the total lipid content in the membranes decreased by half, and the ratio of unsaturated/saturated fatty acids decreased by one-third (Somerville and Browse 1991). Thus, the relationship between the membrane and thermal stability at high temperatures can vary between genotypes and can be used as an important selection criterion for heat stress tolerance. In some species, however, heat tolerance does not correlate with the degree of lipid saturation, suggesting that other factors in addition to membrane thermal stability limit growth at high temperatures.

Proteins are easily denatured by high temperature. As a result, organisms have a set of heat shock proteins (HSPs) that are synthesized in response to high temperature and designed to prevent or reverse the effects of heat on protein denaturation. The expression of HSPs and heat shock factors (proteins that stimulate the expression of HSPs and other heat tolerance genes) increases thermal tolerance in many plants (Sun et al. 2002). Thus, various studies on thermal tolerance have evaluated the expression pattern of HSPs.

High temperatures can cause considerable pre- and post-harvest damages, including leaf, twig and stem burns, leaf senescence and abscission, root growth inhibition and fruit discoloration, and damage, which eventually reduce production. For example, the growth of maize coleoptiles is reduced at 40 °C and ceases at 45 °C (Weaich et al. 1996). In sugarcane grown at high temperatures, the internodes are shorter, and there is greater tillering, early senescence, and a reduction in total biomass (Ebrahim et al. 1998). In rice, high soil temperatures (above 37 °C) between the final tillering stage and panicle initiation reduce the production, filling, and quality of grains (Arai-Sanoh et al. 2010). The development of pollen and anthers is also particularly sensitive to heat (Sato et al. 2006). In tomatoes, microsporogenesis is very sensitive to high temperatures, which can compromise fruiting at the anthesis phase (Sato et al. 2000). In beans, there were substantial reductions in the number of pods and in seed production and a 50 % inhibition of pollen tube viability at temperatures above 37/27 °C day/night (Prasad et al. 2002). During reproduction, a short period of heat stress can cause significant increases in the number and opening of flower buds and in the rate of floral abortion (Young et al. 2004). Furthermore, grain quality in maize is compromised by starch, protein, and oil content reductions (Wilhelm et al. 1999). In wheat, high temperatures during grain filling can change the quality and other physicochemical properties of the flour (Perrotta et al. 1998), including the protein content (Wardlaw et al. 2002). A reduction in the number of grains per spike also occurs at maturity in wheat (Ferris et al. 1998).

The reduction in grain weight in response to heat stress during the initial stage of filling can be partially attributed to the small number of cells in the endosperm, while the subsequent stress reduces starch synthesis due to the reduced availability of assimilates for the grain or direct effects on the synthesis processes in the grains (Yang et al. 2004).

Photosynthesis is limited at temperatures above 35 °C due to the decreased activity of Rubisco. In addition, stomatal closure due to high evaporative air demand, which normally occurs at high temperatures, contributes to the reduction of photosynthesis. Most plants have a considerable capacity to adjust photosynthetic performance to their growth temperatures. Part of that capacity can be assigned to the properties of Rubisco, which is very thermostable. Notably, the rate of carboxylation may increase at 40 °C or above but at a lower speed than the oxygenation rate; consequently, higher photorespiration rates are frequent at high temperatures, which ultimately reduce carbon gain. Therefore, Rubisco that has more carboxylation capacity would be advantageous in situations of excessive heat (Zhu et al. 2010). Regardless, at moderately high temperatures, the low activation state of Rubisco would be the main cause of the decreased photosynthetic rate (Kim and Portis 2005) due to the suppression of Rubisco activase activity (Salvucci and Crafts-Brandner 2004).

The lower regeneration capacity of ribulose-1,5-bisphosphate (RuBP) is another limiting factor for the photosynthetic rate at moderately high temperatures. The decrease in the Rubisco activation state may be a regulatory response to the limitation of one of the processes restricting the regeneration of RuBP, including damage to the thylakoid reactions (Sharkey 2005). When overexpressed, enzymes involved in RuBP regeneration, such as sedoheptulose-1,7-bisphosphatase, reduce the photosynthesis inhibition caused by heat (Feng et al. 2007). In this context, an improved response of photosynthesis to warmer temperatures by manipulating RuBP regeneration seems to be a promising way to increase thermal tolerance.

The accumulation of compatible osmolytes is an important adaptive mechanism to heat stress. Under heat, different plant species can accumulate a variety of osmolytes, such as sugars, sugar alcohols, proline, and betaines, as a way to increase tolerance to heat stress, such as in water and salt stresses. Due to the pivotal role of osmolytes in response to environmental stresses, the selection of materials with a greater potential for the accumulation of these osmolytes, whether through traditional genetic breeding, marker-assisted selection or genetic engineering, is of special interest (Ashraf and Foolad 2007).

In summary, the measurement of membrane thermal stability is a good indicator of high temperature tolerance. In this context, the electrolyte leakage method, which is used to check the loss of cell compartmentalization, seems very sensitive for the quantification of temperature tolerance, as demonstrated in rice (Tripathy et al. 2000) and beans (Thiaw and Hall 2004). The carbon isotope discrimination method in cotton was also positively linked with the selection of more heat-resistant materials (Lu et al. 1996). In addition to these tools, the chlorophyll *a* fluorescence and the SPAD (Soil–Plant Analysis Development) index have been used as good predictors of genotypes that are tolerant to heat stress.

3.4.2 Strategies to Increase Heat Tolerance

To increase heat tolerance, crop breeding has been conducted by biotechnological methods, such as the control of membrane composition (Murakami et al. 2000) or the production of cultivars with the constitutive expression or overexpression of HSPs (Wang and Luthe 2003). Studies have indicated that plant tolerance to heat is a multigenic characteristic. Despite the genetic complexity and difficulties found, some heat-tolerant strains and hybrid cultivars have been developed in crops such as tomato (Scott et al. 1995).

In beans, genes that increase heat tolerance cause a progressive effect on dwarfism due to shortening of the main stalk internodes (Ismail and Hall 1998). Similarly, in cotton, heat-tolerant strains have a substantial reduction in size (Lu and Zeiger 1994). These changes facilitate the planting density and may lead to lower leaf temperatures. High planting density can lead to changes in canopy architecture that can help to reduce thermal stresses with minimal effects on water consumption.

3.5 Nitrogen (N) Deficiency

The availability of N in soils, especially in tropical soils, is usually far below that required for crop species to achieve satisfactory yields. This inadequacy is due to the fundamental importance of N for plant metabolism, as this element is a constituent of chlorophyll, proteins, and nucleic acids, among other molecules, and N is accumulated in large quantities in most species. The major way to overcome N deficiency is through nitrogenous fertilizers, which is applied in large quantities due to low N use efficiency (NUE, i.e., the mass of the harvested product per amount of N applied) of crops, usually approximately 30–40 % (Raun and Johnson 1999). This low NUE implies significant economic and environmental losses. Therefore, NUE optimization in crop species is essential to more sustainable agricultural production.

The search for more efficient cultivars with respect to N use has been the goal of many breeding programs, especially those for cereals because 65 % of the N fertilizer production in the world is used for cereal crops (Garnett et al. 2009). The greatest efforts have been concentrated in rice, maize, and wheat crops (Hirel et al. 2007), for which there is already extensive knowledge about the genetic and physiological determinants of NUE. In this chapter, NUE physiological components, which are the product of the N uptake efficiency (NUpE, i.e., total N in the plant per amount of N applied), and the N utilization efficiency (NUtE, i.e., mass of the harvested product for the total N in the plant), are emphasized.

3.5.1 Physiological Strategies for Increasing NUE

3.5.1.1 Physiological Strategies for Increasing NUpE

The NUpE comprises the uptake of NO_3^- and NH_4^+ ions from the soil, which are held by a series of transporters located in the root epidermis. At the whole plant level, NUpE also depends on the cell storage capacity of these ions and their translocation to the uptake site, which may be in the roots or shoots, depending on the species (Chardon et al. 2010). NO_3^- uptake is performed by high-affinity transporter systems (HATS) or low-affinity transporter systems (LATS), which operate at micromolar and millimolar concentration ranges, respectively. Two gene families encode these transporters: *NRT1*, responsible for HATS, and *NRT2*, for LATS. NH_4^+ uptake also involves high and low-affinity transporters, which are encoded by the *aminomethyltransferase* (*AMT*) gene family. The absorption process is regulated by the ion concentration in the soil and by the N status in the plant. Typically, the level of glutamine acts as a negative signal for absorption. Once absorbed, NO_3^- can be stored in vacuoles, reduced in the root cortex cells, transported to the xylem and then to the shoots, or undergo efflux to the soil. NH_4^+ is almost entirely assimilated in the roots, with very little transported to the aerial parts, but recent studies have suggested that many species can transport considerable amounts of NH_4^+ in the xylem, when the assimilative capacity of the roots has been exceeded (Garnett et al. 2009).

The most direct strategy for increasing NUpE would be an increase in the root-to-shoot ratio, enabling a greater investment in roots and consequently higher nutrient uptake. This strategy, however, also implies carbon deviation from the shoots to the roots, which can limit the carbon fixation ability of the plant as a whole and thus its productivity, as suggested by studies that have identified quantitative trait loci (QTLs) in which there is usually a negative correlation between the size of the root system and the productivity of the plant (Coque and Gallais 2007).

Increased root length density (higher number of roots of a smaller diameter per volume) has been suggested to be a way to increase the specific surface of the root system without altering the proportion of carbon invested. Because N transport occurs primarily through mass flow, however, the root surface may have a limited importance for an increase in N uptake. These aspects are highly dependent on soil type and NO_3^- and NH_4^+ prevalence in the soil. Thus, having a larger root length density may be advantageous in soils with high leaching potential and/or where NH_4^+ (less mobile than NO_3^-) is the predominant form to be absorbed.

The ability of post-anthesis N uptake capacity and its importance in grain filling also plays a key role in crop yield. In maize, 50 % of N is absorbed in the grains during the post-anthesis phase, whereas in rice and wheat, 70–90 % of N is absorbed prior to anthesis (Hirel et al. 2007). Therefore, the selection of genotypes that are able to extend N uptake during the crop cycle can be an important strategy to increase productivity and/or grain quality. In maize, the largest genotypic variation found in protein content breeding was associated with higher N uptake after anthesis (Coque and Gallais 2007). A difficulty often found by crop breeders resides in the negative correlation between productivity and N concentration in grains. In winter wheat, post-anthesis N absorption was shown to be highly correlated with the relationship between protein concentration and productivity; whether this correlation occurs in other crops remains to be demonstrated (Bogard et al. 2010).

At the physiological level, the increase in the number of transporters present in the root would also be a potential alternative for increasing N uptake. Roots usually have a high capacity for NO_3^- absorption, but overexpression of the involved transporters, especially HATS, did not translate into increases in NUE (Good et al. 2004). Alternatively, promising laboratory results have not been validated under field conditions (Quaggiotti et al. 2003). These HATS carriers are subjected to strong transcriptional and post-translational control, which would minimize N uptake during certain stages of the plant cycle, even when the concentration of N in the soil is abundant. Therefore, methods to "circumvent" the regulatory mechanisms of N uptake must be identified. Currently, the overexpression of alanine aminotransferase in the roots has resulted in increased NUpE in canola by reducing the levels of glutamine uptake and increasing the levels of alanine, a metabolite that would be "invisible" to the regulatory systems, allowing increased NO_3^- influx (Good et al. 2007).

In addition to the increase in the NO_3^- influx, an increase in its storage capacity in the cells is also desirable. Although there is some genetic variability for this trait, the extent to which the NUE would be affected by an increased NO_3^- influx remains unclear (Hirel et al. 2007). An increase in N storage capacity should be coupled to a greater ability to use the N excess in the harvested product, if such an increase is a breeding program goal (Triboi and Triboi-Blondel 2002).

3.5.1.2 Physiological Strategies for Increasing NUtE

NUtE is the optimum between two components, the N remobilization efficiency (NRE) and the N assimilation efficiency (NAE). NAE is the assimilation of inorganic N into organic N through a series of enzymatic reactions, which begin in the cytoplasm via the action of nitrate reductase (NR), which reduces NO_3^- to NO_2^-. The NO_2^- is then transported to plastids, where it is rapidly reduced to NH_4^+ by nitrite reductase (NiR), and the NH_4^+ is then assimilated by the glutamine synthetase/2-oxoglutarate aminotransferase (GS/GOGAT) pathway, where GS fixes NH_4^+ into a molecule of glutamate to form glutamine, which further reacts with a molecule of 2-oxoglutarate to form two molecules of glutamate via GOGAT catalysis. NRE is the result of a set of enzymes active during leaf senescence that are responsible for recycling NH_4^+ and amino acids from proteolysis events, and NRE also involves the efficiency of phloem transport of newly synthesized amino acids. The key enzymes involved in N remobilization are cytosolic glutamine synthetase (GS1), glutamate dehydrogenase (GDH), asparagine synthetase (AS), and other transaminases. The importance of these enzymes is due to their products, asparagine and glutamine, which are the main amino acids transported in the phloem during senescence because of their high N/C ratios. NUtE is dependent on the efficiency of N in biomass formation, the effect of N on carbohydrate partition, the nitrate reduction efficiency, and the NRE of senescent tissues (Foulkes et al. 2009). Therefore, approaches to increase NUtE concern both carbon and N metabolism, mainly because the enzymes involved in these routes are co-regulated (Nunes-Nesi et al. 2010), and the existence of overlap in QTLs associated with N assimilation and starch synthesis (Zhang et al. 2010).

For a long time, nitrate reduction was considered a limiting factor in N uptake, leading to great efforts to select cultivars with higher NR activities. In general, this approach resulted in small NUE increments due to the overexpression of NR in transgenic plants (Good et al. 2004). Recent evidence suggests that the limiting factor in nitrate reduction would be providing reducing power in the form of NADH, which makes mitochondrial metabolism a new target to improve NUE (Foyer et al. 2011).

The most promising efforts have been related to the role of enzymes involved in N remobilization due to the uptake of N during grain filling being generally insufficient to meet the demand of the reproductive organs. Among these enzymes, GS1 has received greater attention because of its role in metabolism, in which it participates in all processes involving N uptake and recycling, mainly of NH_4^+ released in the photorespiration process, which can be ten times larger than the NH_4^+ derived from NO_3^- assimilation (Masclaux-Daubresse et al. 2010). The activity of GS1 is considered an excellent marker for an increase in physiological NUE, regardless of the developmental stage, and N nutritional conditions in the plant. The transgenic approaches related to GS1 overexpression are largely positive, but there are insufficient details about plant behavior under field conditions. In rice transgenic lineages overexpressing GS1, for example, there were NUtE increases but not NUE increases under N limiting field conditions with respect to the control plants (Brauer et al. 2011).

The manipulation of photosynthesis at both the leaf and canopy levels is aimed at increasing the photosynthetic N use efficiency (PNUE), i.e., the amount of CO_2 assimilated per unit of N in the leaf. PNUE is also an important strategy for increasing NUtE, especially in species in which assimilated CO_2 has been little explored using breeding strategies. The PNUE depends on nitrogenous compound investment not directly related to the photosynthetic process, such as some secondary metabolic compounds. Reducing the content of N per area unit in parallel to the maintenance of the maximum photosynthetic rate is an important strategy to increase PNUE. In addition, increased solar radiation use efficiency (RUE, i.e., the amount of biomass produced per amount of absorbed solar radiation), through changes in plant architecture that lead to more erect leaves, allows a better irradiance distribution along the canopy, decreasing the canopy extinction coefficient, and increasing the RUE of the whole plant.

Recently, emphasis has been placed on cultivars with a reduced rate of senescence in the field, i.e., cultivars with the "stay-green" characteristic. The functional stay-green phenotypes are those with delayed senescence combined with extended photosynthetic activity, which translates into a greater capacity for the synthesis of photoassimilates throughout the crop cycle and a greater capacity for N uptake post-anthesis. A simple physiological marker for assessing these phenotypes involves the determination of the chlorophyll concentration in leaves, as green leaves generally have higher levels of N. In addition, spectral reflectance methods allow large population screening to identify stay-green phenotypes. Determining whether there is extended photosynthetic activity in these stay-green phenotypes in which the delayed senescence is not always followed by increased photosynthetic activity is, thus, essential. The stay-green phenotypes generally have lower RUEs that may impair the protein content in the grains. Hence, the feasibility of using these phenotypes depends on the characteristics desired in the final product. A point to be considered in this approach is which plant organ contributes the most to the N that is remobilized to the grain. In wheat, the true stalk can account for up to 30 % of remobilized N, and thus, an efficient strategy for wheat breeding would be to increase the stalk capacity to store N coupled to a high efficiency of stalk N remobilization in the seeds, allowing the leaves to remain green longer, contributing a higher supply of assimilates, and leading to a greater extent of N absorption during post-anthesis. Moreover, knowledge of crop physiology per se is crucial in determining the strategy to be adopted in breeding programs aiming to improve the quality of the product harvested because certain features are incompatible with others. For example, when a high content of starch (e.g., maize) or oil (e.g., canola) is desired in the harvestable product, the increased N concentration in seeds varies inversely with the starch and oil levels. Conversely, in rice and wheat, if higher grain protein content is desired, higher N concentration in seeds (or vegetative parts in the case of maize used for silage) becomes the target of the breeder, based on positive correlations between N and protein content in these grains (Chardon et al. 2012).

Given the diversity of strategies aimed at NUE improvement, which may include molecular to macroscopic characteristics, multidisciplinary approaches, including agronomic, genetic, biochemical, and physiological aspects are of special interest. In the short term, the use of appropriate and sustainable agronomic practices is the best approach to increase NUE. One of the factors responsible for low NUE in crops lies in the screening assays conducted under excess N conditions, selecting plants that are less efficient in the use of N. In addition, soil conditions and the environment can be limiting factors for NUE during the crop cycle. Notably, the strategies that plants use to increase NUE under conditions of high N availability may be different from those used during low availability and may also vary between different species.

Improving biological N fixation (BNF) in legumes and the introduction of this mechanism in non-legume crops has been the focus of several studies in recent years, particularly with respect to obtaining satisfactory production and using more sustainable technologies. The main BNF limitations are the high susceptibility of some rhizobia strains to environmental stresses and the low capacity of inoculant strains to compete with the native soil microbial population. One way to overcome these problems is the more careful selection and production of inoculants to be used, isolating specific strains of interest for the species best adapted to the prevailing soil and climate conditions in the local culture (Cummings 2005). The use of transgenic rhizobia strains has also been tested and may be promising way to increase the plant-rhizobia interaction. Attempts to introduce the BNF in non-legumes have been performed using N-fixing endophytic and plant growth-promoting bacteria, especially the diazotrophic bacterium *Gluconacetobacter diazotrophicus*, which colonizes the roots of some tropical grasses, particularly sugarcane but also sweet potatoes, coffee, pineapple, tea, mango, and rice (Deb Roy et al. 2010). Recently, Cocking et al. (2006), using a new inoculation technique, were able to successfully promote the colonization of maize, rice, wheat, and tomato roots with *G. diazotrophicus*, but whether these colonized plants have a better N nutrition under field conditions remains to be determined.

3.6 Phosphorus (P) Deficiency

P is one of 17 essential elements needed for plant growth, playing a central role in a variety of processes, including energy generation, nucleic acid synthesis, membrane synthesis and stability, enzymatic activation and inactivation, redox reactions, signaling, carbohydrate metabolism, and N fixation. Although the total content of P in tropical soils may be relatively high, typically ranging from 500 to 2,000 ppm, the total bioavailable P may be only a few parts per million (Sanyal and Dedatta 1991) due to the adsorption of this element by iron and aluminum (Al) oxyhydroxides, the main constituents of the clay fraction of most tropical soils. Thus, low P availability is a major factor limiting the development and growth of plants. Hereafter, physiological components of P use efficiency (PUE; i.e., mass of harvested product per amount of P applied), which is the product of the P acquisition or uptake efficiency (PUpE; i.e., the total plant P per amount of P applied) and the P utilization efficiency (PUtE; i.e., the mass of harvested product per total amount of P in the plant), are emphasized.

3.6.1 Physiological Strategies for Increasing PUpE

Processes that lead to greater PUpE include changes in the morphology and architecture of the root system, increased production and secretion of phosphatases, increased exudation of organic acids, and increased expression of inorganic phosphate (Pi) transporters (Lynch and Brown 2001). Although outside the scope of this chapter, the most prevalent evolutionary adaptation for terrestrial plants to acquire P occurs through mycorrhizal symbiosis (Tibbett and Sanders 2002).

Adjustments that improve P acquisition in soil are key because of the relative immobility of this element, which has concentrations that are typically high in even the most superficial soil profiles. Root architecture refers to the spatial configuration complexity of the root system, which arises in response to soil conditions. Given the uneven availability of P throughout the soil, root system architecture can improve soil exploitation, especially where mineral elements are more available. Changes in the root system architecture associated with improved P acquisition include basal roots with more horizontal growth, which results in shallower roots, the increased formation of adventitious and lateral roots and the increased density and length of root hairs. Overall, these characteristics were positively related to PUpE in beans (Lynch and Brown 2001; Wang et al. 2010), soybeans (Zhao et al. 2004), and rice (Wissuwa 2003).

Under normal growth and development, plant roots exude a large variety of organic compounds, including simple sugars, organic acids, amino acids, phenolic compounds, quinones, flavonoids, hormones, proteins, and polysaccharides (Marschner 1995). Among these compounds, the organic acids, which favor P solubilization, and strigolactones, which are signaling compounds to mycorrhizal colonization, are key for adaptation to environments with low P availability. Recent studies on the mechanisms of P acquisition show that the plants grown in the presence of low P availability have higher root synthesis of low molecular weight organic acids. In some species, there has been an increase in acid phosphatase activity in roots and in its exudation in response to P deficiency, demonstrating their importance for PUpE increase (Vance et al. 2003). Increased phosphatase activity in the shoots has, however, been found, revealing that the main function of phosphatase appears to be associated with the remobilization of P in the plant as a strategy to keep the level of cytoplasmic Pi adequate to maintaining the metabolism and contributing to the PUtE increase.

3.6.2 Physiological Strategies for Increasing PUtE

Processes that occur using less P and thus promote greater PUtE involve decreased growth rates, greater remobilization of internal Pi, changes in carbohydrate metabolism, avoidance of pathways that require a high supply of P, and the use of alternative respiratory systems (Uhde-Stone et al. 2003). In fact, the greater PUtE is mainly attributed to the efficient reuse and retranslocation of P stored in plants. The highest activity of acid phosphatase contributes to the increased efficiency of P utilization in beans through the remobilization of P from old leaves (Kouas et al. 2009). Studies on sorghum, tomatoes, wheat, and soybeans showed that the remobilization of P is not directly correlated only with senescence, as occurs in young tissues (Akhtar et al. 2008). Another important aspect is the release of Pi from the vacuole when P is limited (Akhtar et al. 2008). Pi is stored in vacuoles when P is abundant and is released under conditions of P deficiency, which helps the plants meet the P demands involved in maintaining Pi homeostasis. The mechanism of transporting P to the vacuole by Pi transporters in the tonoplast is still unclear, and high affinity Pi transporters could play an important role in P uptake (Ai et al. 2009).

The conservation of Pi internal reserves is considered an important adaptation for growth under conditions of low P availability, leading to greater PUtE. Because several enzymes in the glycolytic pathway depend on Pi or adenosines (ATP/ADP) as co-substrates, metabolism could be impaired by reduced Pi availability, as occurs in plants grown in soils with limited P. The energy generation and carbon skeleton production continue under these conditions, as shown by the exudation of substantial amounts of organic acids in the rhizosphere. In response to P limitation, many plants show remarkable metabolic flexibility, using alternative metabolic pathways instead of reactions that are Pi- or ATP-dependent. Although ATP and ADP concentrations are reduced, the pyrophosphate (PPi) concentration appears to be buffered (Duff et al. 1989; Dancer et al. 1990) under Pi deficiency. Considered a by-product of secondary metabolism, PPi can be used as an energy donor, enabling ATP availability for other limiting metabolic processes (Theodorou and Plaxton 1996; Plaxton and Carswell 1999). An example is a well-documented reaction that generates fructose-1,6-bisphosphate, which under P-deficient conditions is catalyzed by the enzyme PPi-dependent phosphofructokinase instead of ATP-dependent phosphofructokinase. Regarding the flexibility of the adaptive response to P deficiency, which is present in the glycolytic pathway, adjustments to maintain mitochondrial electron transport must be considered. The significant reduction of ADP and Pi concentration in plants subjected to P limitation inhibits the cytochrome c pathway because low concentrations of ADP and Pi can result in a high ATP/ADP ratio, which is responsible for the inhibition of oxidative phosphorylation. Thus, the induction of the alternative oxidase pathway during P deficiency has been proposed to be responsible for maintaining mitochondrial electron transport and for metabolic continuity (Vance et al. 2003).

P deficiency usually results in the accumulation of secondary metabolites, such as flavonoids and indole alkaloids. The increased synthesis of anthocyanins is a frequent response of plants subjected to P deficiency, protecting the plant against photo-inhibitory damages (Takahashi et al. 1991). In addition to the involvement of senescence, phenolic compounds may be exudated in the rhizosphere in response to P deficiency, acting as chelating or reducing agents, and increasing the absorption of P. Sclerophylly is another common adaptation concerning the increase of phenolic metabolism in many species in low P availability environments. Sclerophillous species usually have hard and rigid leaves, with heavily lignified secondary walls, which assume a secondary metabolic relationship. Loveless (1962) postulated that the sclerophylly is an adaptation for nutrient storage, particularly of P and N. Unlike primary metabolism, secondary metabolism in general does not consume large amounts of Pi but may recycle significant amounts of Pi from phosphate esters. Secondary metabolism, however, produces excessive reducing equivalents, leading to the acidification of the cytosol, which can lead to the activation of alternative oxidase and other pathways to consume the excess reducing equivalents (Sakano 2001).

3.7 Aluminum (Al) Toxicity

Al is the most abundant metal and the third most common element in the Earth's surface, occurring mainly in the form of stable Al silicate complexes, which is non-toxic to plants (Ma and Ryan 2010). Under acidic conditions, however, the Al solubilizes and forms octahedron hexahydrate ($(Al(H_2O)_6^{3+})$, also known as Al^{3+}. This form of Al is toxic to plants, even at micromolar concentrations (Kochian et al. 2005).

Globally, approximately 30 % of all land area consists of acid soils, and 50 % of the world's cultivable lands are potentially acidic; thus, Al toxicity is one of the most important limitations to agricultural production (Piñeros et al. 2005). In Brazil, for example, more than 500 million hectares are formed of acid soils, especially those covered by savannah (Cerrado biome) vegetation (Vitorello et al. 2005). The soils of these areas have high acidity (average pH 4.6), a high concentration of Al and manganese and deficiencies of Ca^{2+}, Mg^{2+}, and P. These limitations, if not corrected, can lead to remarkable drops in crop productivity. The application of limestone ($CaCO_3 + MgCO_3$) is an effective way to correct soil acidity but is often not an economically viable option for less capitalized producers. In addition, limestone corrects only the surface layers and is not effective in correcting subsoil acidity due to the low mobility of the limestone soluble components.

The development of genotypes tolerant to soil Al has gained high importance. In the last decades, great efforts have been made to identify and characterize the mechanisms of plant tolerance to toxic levels of Al (Samac and Tesfaye 2003). Understanding these mechanisms is key for the development of procedures to rapidly select Al-resistant plants with good performance in acid soils (Barceló and Poschenrieder 2002). There is great variability in Al tolerance between species and even between genotypes within a species (Huang et al. 2009). The mechanisms of Al tolerance can basically be summarized into two classes: (i) those that eliminate absorbed Al or prevent/reduce its uptake by the roots (Al exclusion) and (ii) detoxification mechanisms, which usually act by Al complexation, followed by the transfer, and storage of these complexes in vacuoles (internal tolerance) (Hartwig et al. 2007).

3.7.1 Morpho-Physiological Responses and Alterations to Al Toxicity

Al^{3+} has a high load/atomic radius, which allows it to form highly stable electrostatic bonds with negatively charged compounds, such as phosphates and carboxylic groups (Berthon 1996). Thus, many cellular structures, such as the cell wall, plasma membrane, cytoskeleton, and nucleus, are targets of Al^{3+} toxicity.

The primary site of Al accumulation and toxicity is the root meristem, specifically the distal part of the transition zone. The rapid root growth inhibition after exposure indicates that the Al quickly stops cell expansion and elongation before inhibiting cell division (Kochian et al. 2005). After more prolonged exposure of the root system to Al, its toxicity is manifested through a set of symptoms expressed in its continuous and increasing effect on the morphology and physiology of the roots, involving reductions in the following: biomass; the number and length of the roots, often combined with an increase in the mean radius and root volume; and the uptake of water and mineral nutrients, resulting in severe losses of root elongation and ultimately productivity.

Recent studies have shown that the binding of Al to cell wall components alters the cation exchange capacity (Panda et al. 2009), viscoelasticity (Ma et al. 2004), and other properties of the cell wall, causing changes that block growth. Al can reduce the elasticity of the cell wall and stimulate the synthesis and accumulation of lignin (Peixoto et al. 2007) via the activation of a peroxidase (POD) linked to the cell wall, which is involved in the improvement of hydroxyproline-rich glycoprotein binding to phenolic acids. This improvement then contributes to secondary wall thickening, which results in lower root growth and elongation. Other enzymes activated by Al include NADH oxidase, phenylalanine ammonia-lyase (PAL), and lipoxygenase (LOX). NADH oxidases are responsible for the synthesis of H_2O_2, which is necessary for rapid polymer binding catalyzed by cell wall POD. PAL is a key enzyme in the biosynthesis of phenylpropanoids, and LOX is responsible for the peroxidation of membrane polyunsaturated fatty acids, leading to the formation of hydroperoxides. These compounds are highly reactive and are quickly degraded into compounds that, by the octadecanoic pathway, lead to the production of jasmonic acid, which acts in the lignin synthesis signaling pathway (Xue et al. 2008). Furthermore, Al can disrupt the cytoskeletal dynamics, interacting with microtubules and actin filaments (Kochian et al. 2004). Al can interfere with signal transduction, particularly in the Ca^{2+} signaling pathway (Rengel and Zhang 2003). Al may also increase callose synthesis, blocking the plasmodesmata (Sivaguru et al. 2000) and preventing cell wall loosening, thus limiting the expansion of cells (Jones et al. 2006).

The plasma membrane has a negatively charged surface, making it a sensitive target for Al toxicity. Al strongly binds to phospholipids, which alters the lipid composition (Peixoto et al. 2001), reduces membrane fluidity, and increases the folding density of lipids (Chen et al. 1991). Moreover, Al can inhibit the H^+-ATPase in the plasma membrane, preventing the formation, and maintenance of the H^+ gradient (Ahn et al. 2001). Therefore, Al interferes with the secondary transport of ions, indirectly causing an alteration of ion homeostasis in root cells. Al also rapidly and effectively inhibits the influx of Ca^{2+} into cells by modulating the activity of transporters by changing the membrane potential (Kochian et al. 2005).

Although most of the Al associated with the root system is found in the apoplast (Xue et al. 2008), a significant portion of this cation can penetrate rapidly and interact with molecules and subcellular structures of the symplast (Taylor et al. 2000), such as the nucleus of cells in the meristematic regions of the root apex.

Due to its affinity for phosphate groups, Al binds DNA, negatively affecting its template activity and chromatin structure (Silva et al. 2000) and modifying the cell division process (Barceló and Poschenriede 2002; Kochian et al. 2005).

3.7.2 Physiological Mechanisms of Al Tolerance

Al tolerance or internal detoxification can be accomplished by its complexation in the symplast with different organic compounds and/or by compartmentalization of Al or its complexes in vacuoles (Hartwig et al. 2007). In this context, Al would change little or nothing in plant metabolism, allowing growth and development even after Al input into the symplast. This tolerance mechanism is found mainly in endemic species of regions with acidic soils, where the ability to address Al toxicity is a prerequisite for survival (Ryan and Delhaize 2010). There are, however, few species that accumulate high concentrations of Al in their shoots without suffering from its toxicity (Jansen et al. 2002).

The main tolerance mechanisms that promote Al exclusion or prevent its absorption by the roots include Al immobilization in the cell wall, Al selective permeability in the plasma membrane, pH increases in the rhizosphere or the root apoplast and exudation of organic acids (e.g., citrate, oxalate and malate), and phenolic compounds by the roots. The synthesis and exudation of organic acids is perhaps the major mechanism of Al tolerance. Several evidences support this statement, as discussed by Kochian et al. (2004): (i) a strong correlation exists between Al tolerance and the exudation of organic acids in many species; (ii) the addition of organic acids in the nutritive medium reduces Al toxicity; (iii) Al/organic acid complexes (di-and tri-carboxylic) do not cross the membrane and are not significantly absorbed by the roots; (iv) the exudation of organic acids, activated by Al, occurs at the root apex, the location of the primary effect of Al toxicity; (v) in general, the activation of the exudation mechanism is triggered specifically by Al^{3+}; and (vi) there are, in the plasma membrane, anionic channels activated by Al that facilitate the efflux of organic acids.

Two temporal patterns of organic acid exudation have been identified and characterized in plants (Ma et al. 2001). In Pattern I, the plants are characterized by having an almost immediate response to the release of organic acids by the roots when exposed to Al. This process appears to involve the activation of pre-existing proteins, as found in wheat, tobacco, and barley. In plants with Pattern II (e.g., sorghum), the existence of a lag-phase between Al exposure and organic acid release is found, and this process is assumed to involve the induction of gene expression (Magalhães et al. 2007).

Several studies have shown that excess Al induces ROS production, triggering oxidative stress (Boscolo et al. 2003). Genotypes with more robust antioxidant systems are usually more tolerant to excess Al, but the mechanism by which Al exacerbates the formation of ROS is still not fully understood (Darkó et al. 2004). Although how Al^{3+} acts in the cell to induce ROS formation is not exactly known, its direct involvement in redox reactions seems unlikely because it is not a transition metal. Possibly, Al^{3+}, due to its high affinity for biomembranes, can cause changes in membrane structure, thereby promoting the formation of ROS (Cakmark and Horst 1991). Moreover, the affinity of Al^{3+} for biomembranes can cause rigidity and facilitate chain reactions mediated by Fe^{2+} ions, which increases lipid peroxidation (Yamamoto et al. 2001). As Al accumulates preferentially in the roots, its most important effects occur in this region, and indirect effects are thought to occur in the shoots, mainly by affecting the translocation of nutrients (Lindon et al. 1999). In shoots, however, Al^{3+} can also cause oxidative stress, which can be indirectly monitored by the assessment of fluorescence parameters, particularly those monitoring the maximum quantum yield of photosystem II.

3.7.3 Identification of Tolerant Plants to High Al

The use of Al-tolerant cultivars stands out as the most effective strategy for the production of economically important crops in acid soils. The development of plants resistant to mineral stresses is more profitable than correcting a soil nutrient deficiency or toxicity (Miyasaka et al. 2006). The most used method to evaluate Al toxicity is the comparison of the root growth of plants cultivated in a nutrient solution with an acid pH and Al with the root growth of control plants cultivated in the absence of Al. Therefore, Al differential tolerance may be assessed by comparing the root elongation of different genotypes in solutions with increasing concentrations of Al and is expressed as relative root growth, i.e., the root elongation ratio of the plants from the treatment with Al and those from the treatment without Al. Furthermore, the root growth potential after Al treatment can be evaluated. In this case, the seedlings are grown in a nutrient solution without Al. After approximately 48 h, they are transferred to a nutrient solution containing Al, where they remain for another 48 h, and are then returned to the initial solution for another 72 h. At that point, the seedlings are evaluated for the main root growth after the damage caused by Al^{3+} (Boscolo et al. 2003).

3.8 Conclusion

Over the past years, our understanding of plant adaptation to abiotic stresses has increased enormously. However, there are still many gaps in our knowledge on the physiological mechanisms associated with successful crop growth and production in stressful environments. For instance, drought-stress signaling, which is essential for rational genetic engineering programs aimed at durable stress tolerance, is still very poorly understood (Chinnusamy et al. 2005b). Furthermore, some significant physiological research coping with plant stress tolerance has been conducted using potted plants without the appropriate calibration in the field, which can lead to a waste of time and resources since in most cases results cannot be extrapolated or simulated by crop modeling to describe what may occur under real field conditions. In any case, the advent of new biotechnological tools for proper identification of genetic and physiological determinants involved in plant stress adaptation will certainly contribute to increasing the efficiency of selection for improved crop performance under stress.

References

Ahn SJ, Sivaguru M, Osawa H, Chung GC, Matsumoto H (2001) Aluminum inhibits the H^+-ATPase activity by permanently altering the plasma membrane surface potentials in squash roots. Plant Physiol 126:1381–1390

Ai P, Sun S, Zhao J, Fan X, Xin W, Guo Q, Yu L, Shen Q, Wu P, Miller AJ, Xu G (2009) Two rice phosphate transporters, OsPht1;2 and OsPht1;6, have different functions and kinetic properties in uptake and translocation. Plant J 57:798–809

Akhtar MS, Oki Y, Adachi T (2008) Genetic variability in phosphorus acquisition and utilization efficiency from sparin glysoluble P-sources by *Brassica* cultivars under P-stress environment. J Agron Crop Sci 194:380–392

Arai-Sanoh Y, Ishimaru T, Ohsumi A, Kondo M (2010) Effects of soil temperature on growth and root function in rice. Plant Prod Sci 13:235–242

Ashraf M (2004) Some important physiological selection criteria for salt tolerance in plants. Flora 199:361–376

Ashraf M, Foolad MR (2007) Roles of glycine betaine and proline in improving plant abiotic stress resistance. Environ Exp Bot 59:206–216

Ashraf M, Akram NA (2009) Improving salinity tolerance of plants through conventional breeding and genetic engineering: An analytical comparison. Biotechnol Adv 27:744–752

Ashraf M, Saeed MM, Qureshi MJ (1994) Tolerance to high temperature in cotton (*Gossypium hirsutum* L.) at initial growth stages. Environ Exp Bot 34:275–283

Barceló J, Poschenrieder C (2002) Fast root growth responses, root exudates, and internal detoxification as clues to the mechanisms of aluminium toxicity and resistance: a review. Environ Exp Bot 48:75–92

Barnabás B, Järgen K, Fehér A (2008) The effect of drought and heat stress on reproductive processes in cereals. Plant Cell Environ 31:11–38

Berthon G (1996) Chemical speciation studies in relation to aluminium metabolism and toxicity. Coord Chem Rev 149:241–280

Blum A (2005) Drought resistance, water-use efficiency, and yield potential are they compatible, dissonant, or mutually exclusive? Austr J Agric Res 56:1159–1168

Blumwald E, Aharon GS, Apse MP (2000) Sodium transport in plants. Biochim Biophys Acta 1465:140–151

Bogard M, Allard V, Brancourt-Hulmel M, Heumez E, Machet J, Jeuffroy M, Gate P, Martre P, Gouis JL (2010) Deviation from the grain protein concentration–grain yield negative relationship is highly correlated to post-anthesis N uptake in winter wheat. J Exp Bot 61:4303–4312

Boscolo PRS, Menossi M, Jorge RA (2003) Aluminum-induced oxidative stress in maize. Phytochemistry 62:181–189

Brauer EK, Rochon A, Bi Y, Bozzo GG, Rothstein SJ, Shelp BJ (2011) Reappraisal of nitrogen use efficiency in rice overexpressing glutamine synthetase1. Physiol Plant 141:361–372

Bray EA, Bailey-Serres J, Weretilnyk E (2000) Responses to abiotic stresses. In: Buchanan B, Gruissem W, Jones R (eds) Biochemistry and molecular biology of plants. American Society of Plant Physiologists, Rockville

Cakmak I, Horst WJ (1991) Effect of aluminium on lipid peroxidation, superoxide dismutase, catalase, and peroxidase activities in root tips of soybean (*Glycine max*). Physiol Plant 83:463–468

Chaerle L, Lenk S, Leinonen I, Jones HG, van der Straeten D, Buschnann C (2009) Multi-sensor plant imaging: towards the development of a stress catalogue. Biotechnol J 4:1152–1167

Chardon F, Barthélémy J, Daniel-Vedele F, Masclaux-Daubresse C (2010) Natural variation of nitrate uptake and nitrogen use efficiency in *Arabidopsis thaliana* cultivated with limiting and ample nitrogen supply. J Exp Bot 61:2293–2302

Chardon F, Noël V, Masclaux-Daubresse C (2012) Exploring NUE in crops and in Arabidopsis ideotypes to improve yield and seed quality. J Exp Bot. doi:10.1093/jxb/err353

Chen J, Sucoff EI, Stadelmann EJ (1991) Aluminum and temperature alteration of cell membrane permeability of *Quercus rubra*. Plant Physiol 96:644–649

Chinnusamy V, Jagendorf A, Zhu J (2005a) Understanding and improving salt tolerance in plants. Crop Sci 45:437–448

Chinnusamy V, Xiong L, Zhu J-K (2005b) Use of genetic engineering and molecular biology approaches for crop improvement for stress environments. In: Ashraf M, Harris PJC (eds) Abiotic stresses: plant resistance through breeding and molecular approaches. Food Product Press, New York

Cimato A, Castelli S, Tattini M, Traversi ML (2010) An ecophysiological analysis of salinity tolerance in olive. Environ Exp Bot 68:214–221

Cocking EC, Stone PJ, Davey MR (2006) Intracellular colonization of roots of *Arabidopsis* and crop plants by *Gluconacetobacter diazotrophicus*. In Vitro Cell Dev Biol Plant 42:74–82

Condon AG, Richards RA, Rebetzke GJ, Farquhar GD (2002) Improving intrinsic water-use efficiency and crop yield. Crop Sci 42:122–131

Coque M, Gallais A (2007) Genetic variation for nitrogen remobilization and postsilking nitrogen uptake in maize recombinant inbred lines: heritabilities and correlations among traits. Crop Sci 47:1787–1796

Cummings SP (2005) The role and future potential of nitrogen fixing bacteria to boost productivity in organic and low-input sustainable farming systems. Environ Microbiol 1:1–10

DaMatta FM (2003) Drought as a multidimensional stress affecting photosynthesis in tropical tree crops. In: Hemantaranjan E (ed) Advances in plant physiology, 5th edn. Scientific Publishers, Jodhpur

DaMatta FM, Grandis A, Arenque BC, Buckeridge MS (2010) Impacts of climate changes on crop physiology and food quality. Food Res Int 43:1814–1823

Dancer J, Veith R, Feil R, Komor E, Stitt M (1990) Independent changes of inorganic pyrophosphate and the ATP/ADP or UTP/UDP ratios in plant suspension cultures. Plant Sci 66:59–63

Darkó E, Ambrus H, Stefanovits-Banyai E, Fodor J, Bakos F, Barnabás B (2004) Aluminum toxicity, Al tolerance and oxidative stress in Al-sensitive wheat genotype and in Al-tolerant lines developed by in vitro microspore selection. Plant Sci 166:583–591

Deb Roy B, Deb B, Sharma GD (2010) Role of acetic acid bacteria in biological nitrogen fixation. Biofrontiers 1:47–57

Duff SMG, Moorhead GBG, Lefebvre DD, Plaxton WC (1989) Phosphate starvation inducible 'bypasses' of adenylate and phosphate dependent glycolytic enzymes in *Brassica nigra* suspension cells. Plant Physiol 90:1275–1278

Ebrahim MK, Zingsheim O, El-Shourbagy MN, Moore PH, Komor E (1998) Growth and sugar storage in sugarcane grown at temperature below and above optimum. J Plant Physiol 153:593–602

Ehleringer JR, Hall AE, Farquhar GD (1993) Stable isotopes and plant carbon-water relations. Academic Press, San Diego

FAO (2005) Global network on integrated soil management for sustainable use of salt-affected soils. FAO Land and Plant Nutrition Management Service, Italy

Farquhar GD, Ehleringer JR, Hubick K (1989) Carbon isotope discrimination and photosynthesis. Annu Rev Plant Physiol Plant Mol Biol 40:503–537

Feng LL, Wang K, Li Y, Tan YP, Kong J, Li H, Li YS, Zhu YG (2007) Overexpression of SBPase enhances photosynthesis against high temperature stress in transgenic rice plants. Plant Cell Rep 26:1635–1646

Ferris R, Ellis RH, Wheeler TR, Hadley P (1998) Effect of high temperature stress at anthesis on grain yield and biomass of field grown crops of wheat. Plant Cell Environ 34:67–78

Flowers TJ (2004) Improving crop salt tolerance. J Exp Bot 55:307–319

Foulkes MJ, Hawkesford MJ, Barraclough PB, Holdsworth MJ, Kerr S, Kightley S, Shewry PR (2009) Identifying traits to improve the nitrogen economy of wheat: recent advances and future prospects. Field Crops Res 114:329–342

Foyer C, Noctor G, Hodges M (2011) Respiration and nitrogen assimilation: targeting mitochondria-associated metabolism as a means to enhance nitrogen use efficiency. J Exp Bot 62:1467–1482

Garnett T, Conn V, Kaiser BN (2009) Root based approaches to improving nitrogen use efficiency in plants. Plant Cell Environ 32:1272–1283

Good AG, Shrawat AK, Muench DG (2004) Can less yield more? Is reducing nutrient input into the environment compatible with maintaining crop production? Trends Plant Sci 9:597–605

Good AG, Johnson SJ, De Pauw M, Carroll RT, Savidov N, Vidmar J, Lu Z, Taylor G, Stroeher V (2007) Engineering nitrogen use efficiency with alanine aminotransferase. Can J Bot 85:252–262

Hartwig I, Oliveira AC, Carvalho FIF, Bertan I, Silva JAG, Schmidt DAM, Valério IP, Maia LC, Fonseca DAR, Reis CES (2007) Mecanismos associados à tolerância ao alumínio em plantas. Semina: Ciênc Agr 28:219–228

Hauser F, Horie T (2010) A conserved primary salt tolerance mechanism mediated by HKT transporters: a mechanism for sodium exclusion and maintenance of high K^+/Na^+ ratio in leaves during salinity stress. Plant Cell Environ 33:552–565

Hirel B, Gouis JL, Ney B, Gallais A (2007) The challenge of improving nitrogen use efficiency in crop plants: towards a more central role for genetic variability and quantitative genetics within integrated approaches. J Exp Bot 58:2369–2387

Huang CF, Yamaji N, Mitani N, Yano M, Nagamura Y, Ma JF (2009) A bacterial-type ABC transporter is involved in aluminum tolerance in rice. Plant Cell 21:655–667

IPCC—Intergovernmental Panel on Climate Change (2007) Summary for policymakers of the first volume of "Climate Change 2007". Working Group I of the intergovernmental panel on climate change, Geneva

Ismail AM, Hall AE (1998) Positive and potential negative effects of heat-tolerance genes in cowpea lines. Crop Sci 38:381–390

Ismail AM, Hall AE (1999) Reproductive-stage heat tolerance, leaf membrane thermostability and plant morphology in cowpea. Crop Sci 39:1762–1768

Jacoby RP, Taylor NL, Millar H (2011) The role of mitochondrial respiration in salinity tolerance. Trends Plant Sci 16:614–623

Jansen S, Broadley MR, Robbrecht W, Smets E (2002) Aluminum hyperaccumulation in angiosperms: a review of its phylogenetic significance. Bot Rev 68:235–269

Jones DL, Blancaflor EB, Kochian LV, Gilroy S (2006) Spatial coordination of aluminium uptake, production of reactive oxygen species, callose production and wall rigidification in maize roots. Plant Cell Environ 29:1309–1318

Kim K, Portis ARJR (2005) Temperature dependence of photosynthesis in *Arabidopsis* plants with modifications in Rubisco activase and membrane fluidity. Plant Cell Physiol 46:22–530

Kochian LV, Hoekenga OA, Piñeros MA (2004) How do crop plants tolerate acid soils? Mechanisms of aluminum tolerance and phosphorous efficiency. Annu Rev Plant Biol 55:459–493

Kochian LV, Piñeros MA, Hoekenga OA (2005) The physiology, genetics and molecular biology of plant aluminum resistance and toxicity. Plant Soil 274:75–195

Kouas S, Debe A, Slatni T, Labidi N, Drevon JJ, Abdell C (2009) Root proliferation, proton efflux, and acid phosphatase activity in common bean (*Phaseolus vulgaris*) under phosphorus shortage. J Plant Biol 52:395–402

Läunchli A, Epstein E (1990) Plant responses to saline and sodic conditions. In: Tanji KK (ed) Agricultural salinity assessment and management. American Society of Civil Engineers, New York

Lima ALS, DaMatta FM, Pinheiro HA, Totola MR, Loureiro ME (2002) Photochemical responses and oxidative stress in two clones of *Coffea canephora* under water deficit conditions. Environ Exp Bot 47:239–247

Lindon FC, Ramalho JC, Barreiro MG, Lauriano JA (1999) Modulation of photosystem 2 reactions mediated by aluminium toxicity in *Zea mays*. Photosynthetica 34:151–156
Loveless AR (1962) Further evidence to support a nutritional interpretation of sclerophylly. Ann Bot 26:551–561
Lu Z, Zeiger E (1994) Selection of higher yield and heat resistance in pima cotton has caused genetically determined changes in stomatal conductance. Physiol Plant 92:273–278
Lu Z, Chen J, Percy RG, Sharifi MR, Rundel PW, Zeiger E (1996) Genetic variation in carbon isotope discrimination and its relation to stomatal conductance in pima cotton (*Gossypium barbadense*). Aust J Plant Physiol 23:127–132
Ludlow MM, Santamaria JM, Fukai S (1990) Contribution of osmotic adjustment to grain yield in *Sorghum bicolor* (L.) Moench under water limited conditions. II. Water stress after anthesis. Aust J Agric Res 41:67–78
Lynch JP, Brown KM (2001) Topsoil foraging—an architectural adaptation of plants to low phosphorus. Plant Soil 237:225–237
Ma JF, Ryan PR, Delhaize E (2001) Aluminium tolerance in plants and the complexing role of organic acids. Trends Plant Sci 6:273–278
Ma JF, Nagao S, Sato K, Ito H, Furukawa J, Takeda K (2004) Molecular mapping of a gene responsible for Al-activated secretion of citrate in barley. J Exp Bot 55:1335–1341
Ma JF, Ryan PR (2010) Understanding how plants cope with acid soils. Funct Plant Biol 37:3–6
Magalhães JV, Liu J, Guimarães CT, Lana UGP, Alves VMC, Wang YH, Schaffert RE, Hoekenga OA, Piñeros MA, Shaff JE, Klein PE, Carneiro NP, Coelho CM, Trick HN, Kochian LV (2007) A gene in the multidrug and toxic compound extrusion (MATE) family confers aluminum tolerance in sorghum. Nat Genet 39:1156–1161
Marcum KB (1998) Cell membrane thermostability and whole plant heat tolerance of Kentucky bluegrass. Crop Sci 38:1214–1218
Marschner H (1995) Mineral nutrition of higher plants, 2nd edn. Academic Press, Boston
Masclaux-Daubresse C, Daniel-Vedele F, Dechorgnat J, Chardon F, Gaufichon L, Suzuki A (2010) Nitrogen uptake, assimilation and remobilization in plants: challenges for sustainable and productive agriculture. Ann Bot 105:1141–1157
Miyasaka SC, Hue NV, Dunn MA (2006) Aluminun. In: Barker AV, Pilbeam DJ (eds) Handbook of plant nutrition. CRC Press, New York
Moinuddin KR (2004) Osmotic adjustment in chickpea in relation to seed yield and yield parameters. Crop Sci 44:449–455
Moller IS, Tester M (2007) Salinity tolerance of Arabidopsis: a good model for cereals? Trends Plant Sci 12:534–540
Monneveux P, Sheshshayee MS, Akhter J, Ribaut JM (2007) Using carbon isotope discrimination to select maize (*Zea mays* L.) inbred lines and hybrids for drought tolerance. Plant Sci 173:390–396
Morgan JM (1992) Osmotic components and properties associated with genotypic differences in osmoregulation in wheat. Aust J Plant Physiol 19:67–76
Morison JIL, Lawlor DW (1999) Interactions between increasing CO_2 concentration and temperature on plant growth. Plant Cell Environ 22:659–682
Munns R, Tester M (2008) Mechanisms of salinity tolerance. Annu Rev Plant Biol 59:651–681
Munns R, James RA, Sirault XRR, Furbank RT, Jones HG (2010) New phenotyping methods for screening wheat and barley for beneficial responses to water deficit. J Exp Bot 61:3499–3507
Murakami Y, Tsuyama M, Kobayashi Y, Kodama H, Iba K (2000) Trienoic fatty acids and plant tolerance of high temperature. Science 287:476–479
Nunes-Nesi A, Fernie AR, Stitt M (2010) Metabolic and signaling aspects underpinning the regulation of plant carbon nitrogen interactions. Mol Plant 3:973–996
Panda SK, Baluska F, Matsumoto H (2009) Aluminium stress signaling in plants. Plant Signal Behav 4:592–597
Parry MAJ, Flexas J, Medrano H (2005) Prospects for crop production under drought: research priorities and future directions. Ann App Biol 147:211–226

Passioura JB (1977) Grain yield, harvest index and water use of wheat. J Aust Inst Agric Sci 43:117–120
Passioura JB (1997) Drought and drought tolerance. In: Belhassen E (ed) Drought tolerance in higher plants: genetical, physiological and molecular biological analysis. Kluwer Academic Publishers, Dordrecht
Peet MM, Willits DH (1998) The effect of night temperature on greenhouse grown tomato yields in warm climate. Agric Forest Meteorol 92:191–202
Peixoto PHP, Cambraia J, Sant'anna R, Mosquim PR, Moreira MA (2001) Aluminum effects on fatty acid composition and lipid peroxidation of a purified plasma membrane fraction of root apices of two sorghum cultivars. J Plant Nutr 24:1061–1070
Peixoto PHP, Pimenta DL, Cambraia J (2007) Alterações morfológicas e acúmulo de compostos fenólicos em plantas de sorgo sob estresse de alumínio. Bragantia 66:17–25
Perrotta C, Treglia AS, Mita G, Giangrande E, Rampino P, Ronga G, Spano G, Marmiroli N (1998) Analysis of mRNAs from ripening wheat seeds: the effect of high temperature. J Cereal Sci 27:127–132
Piñeros MA, Shaff JE, Manslank HS, Alves VMC, Kochian LV (2005) Aluminum resistance in maize cannot be solely explained by root organic acid exudation. A comparative physiological study. Plant Physiol 137:231–241
Pinheiro HA, DaMatta FM, Chaves ARM, Loureiro ME, Ducatti C (2005) Drought tolerance is associated with root depth and stomatal control of water use in clones of *Coffea canephora*. Ann Bot 96:101–108
Plaxton WC, Carswell MC (1999) Metabolic aspects of the phosphate starvation response in plants. In: Lerner HR (ed) Plant responses to environmental stresses: from phytohormones to genome reorganization. Marcel-Dekker, New York
Posch S, Bennett LT (2009) Photosynthesis, photochemistry and antioxidative defence in response to two drought severities and with re-watering in *Allocasuarina luehmannii*. Plant Biol 11:83–93
Prasad PVV, Boote KJ, Allen LH, Thomas JMG (2002) Effects of elevated temperature and carbon dioxide on seed-set and yield of kidney bean (*Phaseolus vulgaris* L.). Global Change Biol 8:710–721
Quaggiotti S, Ruperti B, Borsa P, Destro T, Malagoli M (2003) Expression of a putative high-affinity NO_3^- transporter and of an H^+-ATPase in relation to whole plant nitrate transport physiology in two maize genotypes differently responsive to low nitrogen availability. J Exp Bot 54:1023–1031
Raun WR, Johnson GV (1999) Improving nitrogen use efficiency for cereal production. Agron J 91:357–363
Rengel Z, Zhang W (2003) Role of dynamics of intracellular calcium in aluminium-toxicity syndrome. New Phytol 159:295–314
Ryan PR, Delhaize E (2010) The convergent evolution of aluminium resistance in plants exploits a convenient currency. Funct Plant Biol 37:275–284
Sakano K (2001) Metabolic regulation of pH in plant cells: role of cytoplasmic pH in defense reaction and secondary metabolism. Int Rev Cytol 206:1–44
Salvucci ME, Crafts-Brandner SJ (2004) Relationship between the heat tolerance of photosynthesis and the thermal stability of Rubisco activase in plants from contrasting thermal environments. Plant Physiol 134:1460–1470
Samac DA, Tesfaye M (2003) Plant improvement for tolerance to aluminum in acid soils—a review. Plant Cell Tissue Organ Cult 75:189–207
Sambatti JB, Caylor KK (2007) When is breeding for drought tolerance optimal if drought is random? New Phytol 175:70–80
Sanyal SK, DeDatta SK (1991) Chemistry of phosphorus transformations in soil. Adv Soil Sci 16:1–120
Sato S, Peet MM, Thomas JF (2000) Physiological factors limit fruit set of tomato (*Lycopersicon esculentum* Mill) under chronic, mild heat stress. Plant Cell Environ 23:719–726

Sato S, Kamiyamam M, Iwata T, Makita N, Furukawa H, Keda I (2006) Moderate increase of mean daily temperature adversely affects fruit set of *Lycopersicon esculentum* by disrupting specific physiological processes in male reproductive development. Ann Bot 97:731–738

Scott JW, Olson SM, Howe TK, Stoffella PJ, Bartz JA, Bryan HH (1995) 'Equinox' heat-tolerant hybrid tomato. Hortic Sci 30:647–648

Sharkey TD (2005) Effects of moderate heat stress on photosynthesis: importance of thylakoid reactions, rubisco deactivation, reactive oxygen species, and thermotolerance provided by isoprene. Plant Cell Environ 28:269–277

Silva IR, Smyth TJ, Moxley DF, Carter TE, Allen NS, Rufty TW (2000) Aluminum accumulation at nuclei of cells in the root tip. Fluorescence detection using lumogallion and confocal laser scanning microscopy. Plant Physiol 23:543–552

Singsaas EL, Laporte MM, Shi JZ, Monson RK, Bowling DR, Johnson K, Lerdau M, Jasentuliyana A, Sharkey TD (1999) Leaf temperature fluctuation affects isoprene emission from red oak (*Quercus rubra*) leaves. Tree Physiol 19:917–924

Sivaguru M, Fujiwara T, Samaj J, Baluska F, Yang Z, Osawa H, Maeda T, Mori T, Volkmann D, Matsumoto H (2000) Aluminum-induced 1,3-β-D-glucan inhibits cell-to-cell trafficking of molecules through plasmodesmata: a new mechanism of aluminum toxicity in plants. Plant Physiol 124:991–1005

Somerville C, Browse J (1991) Plant lipids, metabolism and membranes. Science 252:80–87

Sun W, van Montagu M, Verbruggen N (2002) Small heat shock proteins and stress tolerance in plants. Biochim Biophys Acta 1577:1–9

Takahashi A, Takeda K, Ohnishi T (1991) Light-induced anthocyanin reduces the extent of damage to DNA in UV-irradiated *Centaura cyanus* cells in culture. Plant Cell Physiol 32:541–547

Tambussi EA, Bort J, Araus JL (2007) Water use efficiency in C_3 cereals under Mediterranean conditions: a review of physiological aspects. Ann Appl Biol 150:307–321

Tardieu F (2005) Plant tolerance to water deficit: physical limits and possibilities for progress. Comptes Rendus GeoSci 337:57–67

Taylor GJ, Mcdonald-Stephens JL, Hunter DB, Bertsch PM, Elmore D, Rengel Z, Reid RJ (2000) Direct measurement of aluminum uptake and distribution in single cells of *Chara corallina.* Plant Physiol 123:987–996

Tester N, Davenport R (2003) Na^+ tolerance and Na^+ transport in higher plants. Ann Bot 91:1–25

Theodorou ME, Plaxton WC (1996) Purification and characterization of pyrophosphate-dependent phosphofructokinase from phosphate-starved *Brassica nigra* suspension cells. Plant Physiol 112:343–351

Thiaw S, Hall AE (2004) Comparison of selection for either leaf-electrolyte-leakage or pod set in enhancing heat tolerance and grain yield of cowpea. Field Crops Res 86:239–253

Thomson WW, Faraday CD, Oros JW (1988) Salt glands. In: Baker DA, Hall JL (eds) Solute transport in plant cells and tissues. Longman Scientific and Technical, Harlow

Tibbett M, Sanders FE (2002) Ectomycorrhizal symbiosis can enhance plant nutrition through improved access to discrete organic nutrient patches of high resource quality. Ann Bot 89:783–789

Triboi E, Triboi-Blondel A (2002) Productivity and grain or seed composition: a new approach to an old problem. Eur J Agron 16:163–186

Tripathy JN, Zhang J, Robin S, Nguyen TT, Nguyen HT (2000) QTLs for cell membrane stability mapped in rice under drought stress. Theor Appl Genet 100:1197–1202

Tuberosa R, Salvi S (2006) Genomics-based approaches to improve drought tolerance of crops. Trends Plant Sci 11:405–412

Türkan I, Demiral T (2009) Recent developments in understanding salinity tolerance. Environ Exp Bot 67:2–9

Uhde-Stone C, Zinn KE, Ramirez-Yañez M, Li A, Vance CP, Allan DL (2003) Nylon filter arrays reveal differential gene expression in proteoid roots of white lupin in response to phosphorus deficiency. Plant Physiol 131:1064–1079

Vance CP, Uhde-Stone C, Allan DL (2003) Phosphorus acquisition and use: critical adaptations by plants for securing a nonrenewable resource. New Phytol 157:423–447
Vitorello VA, Capaldi FR, Stefanuto VA (2005) Recent advances in aluminium toxicity and resistance in higher plants. Braz J Plant Physiol 17:129–143
Wahid A, Shabbir A (2005) Induction of heat stress tolerance in barley seedlings by pre-sowing seed treatment with glycinebetaine. Plant Growth Regul 46:133–141
Wang D, Luthe DS (2003) Heat sensitivity in a bentgrass variant. Failure to accumulate a chloroplast heat shock protein isoform implicated in heat tolerance. Plant Physiol 133:319–327
Wang X, Shen J, Liao H (2010) Acquisition or utilization, which is more critical for enhancing phosphorus efficiency in modern crops? Plant Sci 179:302–306
Wardlaw IF, Blumenthal C, Larroque O, Wrigley CW (2002) Contrasting effects of chronic heat stress and heat shock on kernel weight and flour quality in wheat. Funct Plant Biol 29:25–34
Weaich K, Briston KL, Cass A (1996) Modeling preemergent maize shoot growth. II. High temperature stress conditions. Agron J 88:398–403
Wilhelm EP, Mullen RE, Keeling PL, Singletary GW (1999) Heat stress during grain filling in maize: effects on kernel growth and metabolism. Crop Sci 39:1733–1741
Wissuwa M (2003) How do plants achieve tolerance to phosphorus deficiency? Small causes with big effects. Plant Physiol 133:1947–1958
Xue YJ, Tao L, Yang ZM (2008) Aluminum-induced cell wall peroxidase activity and lignin synthesis are differentially regulated by jasmonate and nitric oxide. J Agric Food Chem 56:9676–9684
Yamamoto Y, Kobayashi Y, Matsumoto H (2001) Lipid peroxidation is an early symptom triggered by aluminum, but not the primary cause of elongation inhibition in pea roots. Plant Physiol 125:199–208
Yang JC, Zhang JH, Wang ZQ, Xu GW, Zhu QS (2004) Activities of key enzymes in sucrose-to-starch conversion in wheat grains subjected to water deficit during grain filling. Plant Physiol 135:1621–1629
Young LW, Wilen RW, Bonham-Smith PC (2004) High temperature stress of *Brassica napus* during flowering reduces micro- and megagametophyte fertility, induces fruit abortion, and disrupts seed production. J Exp Bot 55:485–495
Zhang N, Gibon Y, Gur A, Chen C, Lepak N, Hoehne M, Zhang Z, Kroon D, Tschoep H, Stitt M, Buckler E (2010) Fine quantitative trait loci mapping of carbon and nitrogen metabolism enzyme activities and seedling biomass in the maize IBM mapping population. Plant Physiol 154:1753–1765
Zhao J, Fu J, Liao H, He Y, Nian H, Hu Y, Qiu L, Dong Y, Yan X (2004) Characterization of root architecture in an applied core collection for phosphorus efficiency of soybean germplasm. Chin Sci Bull 49:1611–1620
Zhu X, Long SP, Ort DR (2010) Improving photosynthetic efficiency for greater yield. Annu Rev Plant Biol 61:235–261

Chapter 4
Breeding for Nitrogen Use Efficiency

**Júlio César DoVale, Rodrigo Oliveira DeLima
and Roberto Fritsche-Neto**

Abstract Nitrogen (N) is a constituent of several organic compounds in plants. Consequently, this element is the most essential to plants. The identification of genotypes able to use nitrogen more efficiently is a slow process due the complexity of the metabolism of nitrogen, the influence of several environmental factors‘ and the great variation in the species studied. N use efficiency (NUE) is defined as the weight of the gains divided by the amount of N available in the soil. Because NUE involves several components, there are several ways to modify gene expression and thereby increase the NUE, from classic plant breeding methods to applied biotechnology. Strategies to improve the absorption and assimilation of N are currently being evaluated in many plant species under field conditions and should be implemented quickly to breeding strategies. While methods for understanding the mechanisms of remobilization and utilization of N during grain development are still not defined, once discovered, they will significantly increase NUE in crop species.

Keywords Plant breeding · Nutrition stress · Biotechnology · Quantitative genetics

J. C. DoVale · R. O. DeLima · R. Fritsche-Neto (✉)
Department of Plant Science, Federal University of Viçosa,
Vicosa-MG, Brazil
e-mail: roberto.neto@ufv.br

J. C. DoVale
e-mail: julio.vale@ufv.br

R. O. DeLima
e-mail: rodrigoodema@gmail.com

R. Fritsche-Neto and A. Borém (eds.), *Plant Breeding for Abiotic Stress Tolerance*,
DOI: 10.1007/978-3-642-30553-5_4, © Springer-Verlag Berlin Heidelberg 2012

4.1 Introduction

Nitrogen (N) is a constituent of several organic compounds in plants, including amino acids, proteins, enzymes, and chlorophyll. Consequently, the primary biochemical reactions that occur in plants involve nitrogen, and thus this element is the most essential to plants. Proof of this nitrogen requirement was demonstrated by the increased productivity of crops grown in the last century, which was strongly associated with increased N fertilization (Hirel et al. 2007).

Because of the high cost of nitrogen fertilizers, in many situations, nitrogen use may not have enabled the increase of the production for small farms, especially in marginal regions of cultivation. However, in developed countries, where higher amounts of N were applied to the soil, the overuse of this nutrient has caused health and environmental problems due to pollution (Algren et al. 2008).

Future challenges include meeting the increasing needs of the world population, increasing food security in regions where there is hunger and improving the use of non-renewable resources. These challenges could be accomplished through the development of highly productive and sustainable agriculture. However, to achieve these objectives, nitrogen must be efficiently used for the development of cultivars.

Moll et al. (1982) defined N use efficiency (NUE) as the weight of the gains divided by the amount of N available in the soil. They defined two primary components of N use efficiency: N uptake efficiency (NUpE) and N utilization efficiency (NUtE). The NUpE is defined as the total amount of the N applied to soil. The variable NUtE is obtained by the ratio between grain weight and the total amount of N in the mature plant. Using these two primary components, we can obtain a value for NUE using the product of NUpE and NUtE.

The identification of genotypes able to use nitrogen more efficiently is a slow process due the complexity of the metabolism of nitrogen, the influence of several environmental factors (Machado et al. 1992), and the great variation in the species studied. Therefore, in this chapter, we will focus on maize, which has been well studied and has strong requirements for nitrogen for the cultivation of this species.

4.2 Germplasms and Genetic Variability

Although it is known that the genetic variability of the maize germplasm is due to low levels of nitrogen present in the soil, the physiological and genetic bases of this variability have not been fully elucidated (Lemaire et al. 1996). In some species, this genetic variability might confer the ability to store large amounts of N during periods of abundant supply, which could be assimilated during periods of scarcity. However, an adequate supply of nitrogen is not always provided.

Every year, the CIMMYT (International Maize and Wheat Improvement Center) evaluates the tropical and temperate maize germplasm for NUE in Mexico

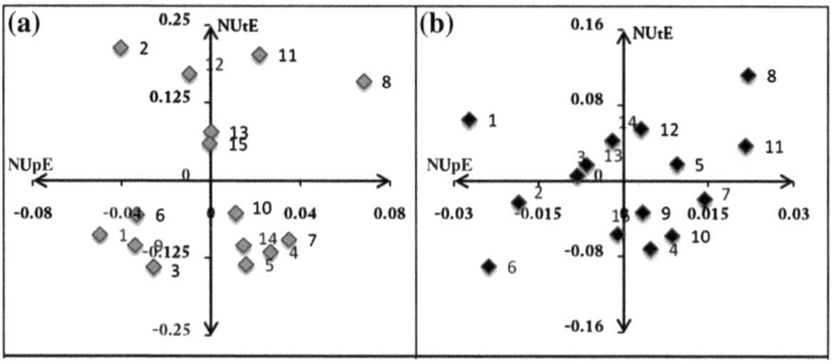

Fig. 4.1 Dispersion of the genotypic values for NUtE and NUpE under high (**a**) and low (**b**) N conditions in tropical maize inbred lines. *Source* Adapted from DoVale (2011)

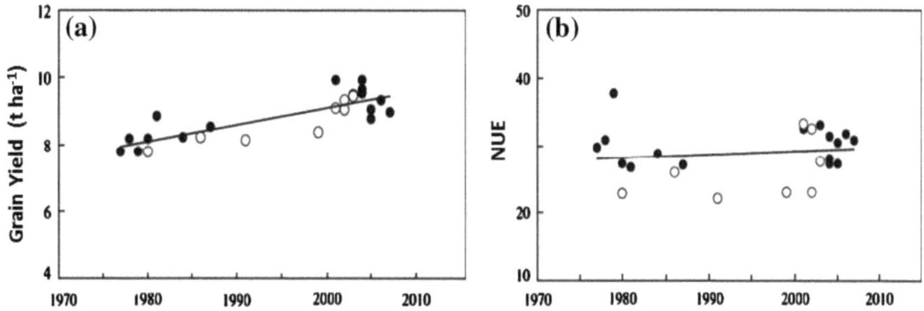

Fig. 4.2 **a** Grain yield and **b** NUE in 39 experimental trials with wheat cultivars evaluated from 2003 to 2007. *Source* Adapted from Sylvester-Bradley and Kindred (2009)

and, most recently, Zimbabwe and Kenya. However, on average, only 3 % of the genotypes show potential for exploitation under low N availability in the soil (Bänzinger et al. 2000).

In Brazil, Fidelis et al. (2007) observed an increased genetic variability in NUE among local populations of open-pollinated tropical maize, with values of grain yield ranging from 2.0 to 4.0 t ha^{-1}. Despite the low grain yield compared to that obtained under ideal growing conditions, the authors concluded that the selection of genotypes provides enough genetic variability to be used by small- and medium-sized farms from marginal areas.

To identify genetic variability in NUE, DoVale (2011) plotted the genotypic values observed in tropical maize lines under high (Fig. 4.1a) and low N conditions (Fig. 4.2b). The author found dispersion in four quadrants: IQ and QII are in the upper right and left quadrants, respectively, and QIII and QI are the lower left and right quadrants, respectively. The Q is characterized by positive NUE values, and the converse is observed in QIII (Fig. 4.1). This methodology reveals that there is a contrasting performance and genetic variability among maize lines for NUE. Consequently, it is possible to select maize with improved NUE values.

Although studies have demonstrated that there is genetic variability for most agronomic traits under low nitrogen conditions, in these environments, there is a reduction in the genetic variability because these genotypes tend to have similar performances (Gallais and Hirel 2004). Moreover, under low N, the means and the heritabilities are smaller, and the coefficients of experimental variation are greater compared with those under ideal or high N conditions. Consequently, selection is more difficult under low N than in ideal conditions. However, using a temperate germplasm, Presterl et al. (2003) have demonstrated that it is possible to select maize genotypes under low N conditions, even with significant reduction in the grain yield.

Genetic variability is manifested differently in the contrasting environments for N availability in the soil. The direct selection under low N availability contributes more effectively to the NUE than the indirect selection under normal N conditions. For example, currently, wheat inbred lines are being directly selected under low N availability, accelerating the generation of genotypes that are adapted to these conditions (Laperche et al. 2006).

Modern cultivars are developed under conditions of high soil fertility in an effort to increase productivity. Thus, it is likely that the selection of genotypes that are more responsive to fertilization has led to the reduction of the allelic diversity associated with tolerance to low N conditions (Wissuwa et al. 2009). Therefore, the germplasm selected under optimal conditions of cultivation might not be suitable for use under abiotic stress conditions (Souza et al. 2009) because the genes that control the grain yield under abiotic stress conditions are different from those that control the grain yield under optimal conditions of cultivation (Souza et al. 2008).

For example, in relation to the increase of the NUE (Fig. 4.2b), the grain yield increase in wheat cultivars (Fig. 4.2a) has remained essentially unchanged in the last 40 years of improvement.

4.3 Inheritance, Maternal Effects and Relationship Among Traits

To understand the genetic control of NUE, studies concerning inheritance and its relationship to the components of NUE and productivity have been conducted annually. The main objective of this research is to determine the most appropriate selection strategy and the main traits that should be considered in a program of plant breeding for NUE. The genetic control of NUE is complex and polygenic (Loqué and Von wirén 2004). Using data from several trials, Presterl et al. (2003) estimated the heritability coefficients for grain yield in maize under ideal and low N conditions. The values obtained ranged from 36 to 96 % under ideal conditions and from 41 to 88 % under low N conditions.

The levels of N availability in the soil and the studied germplasm influence genetic control and the additive and non-additive genetics effects of most traits (Chun et al. 2005). For example, under low N availability, Souza et al. (2008) observed that the additive genetic effects were more important for NUE in maize than the non-additive effects. Consequently, the selection could be performed using inbred lines because the hybrids obtained from these crosses would also be NUE. However, DeLima (2010) observed the opposite effect, i.e., non-additive genetic effects were more important for the expression of NUE in maize than the additive effects. Thus, the performance of the hybrids must be considered while selecting superior genotypes.

Notably, in addition to nuclear genetic effects, the inheritance and expression of a character depends on the maternal effect, i.e., the contribution of the female parent to the genotype of their offspring. Consequently, if there is a significant maternal effect, there will be differences in the choice of the female parent to use for a particular crossing.

Several quantitative and molecular genetics studies have shown that cytoplasmic factors contribute significantly to the variation and inheritance of quantitative and qualitative characters in plants (Roach and Wulff 1987). Thus, a quantitative analysis of the maternal effect using reciprocal crosses is a strategy that might increase the efficiency of breeding for NUE because physiological parameters, such as photosynthetic efficiency, CO_2 exchange, and energy production (via the mitochondria), might be related to NUE. Miranda et al. (2005) used this strategy to select maize inbred lines with increased N use efficiencies based on the chlorophyll meter readings during the flowering of the plant.

Several secondary traits are related to NUE and have been used to increase the efficiency of the selection process and, consequently, support the development of cultivars that produce satisfactorily under low N availability. Among these traits, an increase in enzyme activity (Chevalier and Schrader 1977) has been shown to affect the components of NUE (Moll et al. 1982) as well as the architecture and morphology of the root (Chun et al. 2005).

In relation to the components of the NUE, Gallais and Hirel (2004) noted that the efficiency of N utilization was significantly negatively correlated with the N uptake efficiency ($r_p = -0.80$), which indicates an antagonistic effect between these NUE components that is independent of the availability of N. There are three hypotheses that could explain this phenomenon: (i) the activities of the N transporters and enzymes involved in the assimilation of this nutrient occur at different proportions; (ii) there is degradation of the leaf proteins, particularly rubisco that contributes to the yield; and (iii) the N utilization is most pronounced where the absorption is lower or stopped. According to the authors, under ideal N conditions in the soil, the genetic variability for NUE is a function of the difference in the NUpE, and under low N availability, it is a function of the NUtE. However, Le Gouis et al. (2000) observed the opposite effect, i.e., the NUpE explains most of the genetic variability of the NUE under low N availability in the soil.

Studies concerning root morphology traits will ascertain relevant information regarding NUE and its components. DeLima (2010) studied the correlation between root morphology traits and the components of NUE in tropical maize (Table 4.1).

Under the two N conditions studied, all estimates of the phenotypic correlation ranged from medium to low. However, these estimates reaffirm the results obtained by Gallais and Hirel (2004), where NUtE was more important for NUE

Table 4.1 Estimates of the simple correlation between the lateral (RL_{LAT}, *m*) and axial (RL_{AXI}, *m*) root lengths as well as the N uptake (NUpE), utilization (NUtE), and use (NUE) efficiencies in 188 maize genotypes evaluated under contrasting N conditions

Traits	NUpE[a]	NUpE[b]	NUtE[a]	NUtE[b]	NUE[a]	NUE[b]
RL_{LAT}	0.36**	0.10[c]	−0.07[c]	0.47**	0.32**	0.27**
RL_{AXI}	0.27**	−0.04[c]	−0.10[c]	0.49**	0.29**	0.16*

* $p \leq 0.05$; ** $p \leq 0.01$, respectively

[a] Estimates of correlations obtained under ideal

[b] Low N conditions

[c] Non-significant. *Source* Adapted from DeLima(2010)

Table 4.2 Indirect gains for N use (NUE), N uptake (NUpE), and N utilization (NUtE) efficiencies based on selection according to the activity of glutamine synthetase (GS) in tropical maize under low N conditions

Indirect gains (%)	NUE	NUpE	NUtE
Glutamine synthetase	24.93	21.16	24.01

Source Adapted from Oliveira (2009)

than NUpE under low N availability, and under ideal N conditions, the NUpE was the most important component. Nevertheless, the root traits do not explain the difference among the genotypes for NUE. These results contradict those obtained by Chun et al. (2005), where the root length was important for NUE under low N conditions.

It has been suggested that the NUtE is associated with enzymes involved in the metabolism of the N. Nitrate reductase (NR) and glutamine synthetase (GS) are key enzymes in the reduction and incorporation of N in organic compounds. Machado (1997) evaluated tropical maize inbred lines under different N conditions and found that these enzymes were effective in discriminating the inbred lines to the NUE. In another study, Gallais and Hirel (2004) observed that under low N conditions, the GS activity in the vegetative stage of plant development was positively correlated with grain yield.

Considering the possibility of indirect selection for NUE and its components, Oliveira (2009) estimated the gains observed were due to the GS activity in tropical maize inbred lines under low N availability. For all traits, the gains observed were positive and of high magnitude (Table 4.2). Consequently, the GS has potential for use in the selection and development of maize cultivars with high NUE.

4.4 Induction of Stress: Duration, Intensity and Uniformity

Stress management is crucial for successful breeding programs under abiotic stress conditions. Under low N availability, if the induction of stress is too severe, as previously mentioned, it could “overshadow” the genetic variability and thus make the selection impractical.

Several studies have reported the selection of genotypes with more efficient N use. Most of the work was conducted under field conditions, although some researchers evaluated the genotypes in the greenhouse. However, in both cases, the goal was the same: to clearly simulate the conditions of low N availability under actual cultivation conditions.

Duration, intensity, and uniformity are factors that should be considered in the management of appropriate stress (Bänzinger et al. 2000). The duration should be long enough to correlate the stress with the critical stages of plant growth and involve traits related to improved nutritional efficiency. The intensity must be severe enough to affect traits important to productivity, i.e., to allow differentiation in the germplasm when plants are subjected to ideal N conditions. Finally, the management should be spatially and temporally uniform to easily observe the potential genetic variance and the genetic gains.

The amount of N used to characterize environments with low N availability is a function of the productivity obtained under these conditions. In maize, the productivity under N stress is generally 25–35 % lower than the productivity obtained under ideal N conditions (Bänzinger et al. 1997). Consequently, 20 and 25 % of the amount of N must be used under ideal conditions to simulate the stress of N. Thus, if an average yield of 7 t ha^{-1} using 200 kg N ha^{-1} is expected under ideal growing conditions, under stress conditions, the average grain yield is reduced to 1.5–2.5 t ha^{-1}, and 40–50 kg ha^{-1} N should be used. The indication of these amounts of N are consistent with those used in the experiments under N stress in the Maize Program from the Federal University of Viçosa, where researchers have been working for 10 years to improve the management of abiotic stress. Typically, 250 kg ha^{-1} of 04-14-08 are planted for experiments, and at fourth leaf stages, 20 kg ha^{-1} of N is used (Fidelis et al. 2005).

For the experiments conducted in the greenhouse, opaque pots filled with inert substrate are typically used. The N is supplied using a nutrient solution every second day, from the seventh day after seedling emergence. Normally, modified Hoagland solution containing two contrasting concentrations of N (10 mM for ideal growing conditions and 1 mM for low N availability) is used (Oliveira 2009).

4.5 Strategies for Selection and Breeding

The selection of traits with increased heritability and additive genetic control can be achieved using the individual performance, i.e., the performance of the inbred lines or of the populations. When the trait is of low heritability or presents inheritance due to non-additive genetic effects, the selection of superior genotypes must be based on the performance of the hybrids. Moreover, if there is a significant maternal effect, as previously mentioned, there will be differences in the choice of the female parent to use in a particular crossing.

As demonstrated in this chapter, in the improvement for NUE, the results are contradictory depending on the N levels and germplasm used. Thus, it is essential in each breeding program to determine the inheritance of NUE in the germplasm studied. The identification and quantification of genetic correlations between the nutritional efficiency and agronomic traits are also important to help in the selection. Therefore, the aim is to determine whether it is possible to indirectly select for traits with low heritability, which are typically difficult to evaluate because of the experimental difficulties and/or high cost of measurement.

Finally, it is important to quantify the correlation between the genotype and concentration of N available. If the correlation is significant to the selection of traits, then the selection must not be based on the performance of genotypes in only one environment because many genes are expressed and function differently depending on the availability of N in the soil.

The breeding methods for NUE are chosen as a function of the reproductive system of the species, desired cultivar type and the heritability and genetic control from more important traits.

The population methods are based on recurrent selection because they aim to increase the frequency of favorable alleles for quantitative traits through cycles of repeated selection without significantly reducing the genetic variability of the population. The recurrent selection can divided into three: obtaining progeny, the evaluation of progeny and the recombination of selected progeny (Comstock et al. 1949). The improved populations can be used repeatedly to start a new cycle of recurrent selection after recombination of the higher progenies (Bernardo 2002).

When NUE is influence by additive genetic control, intrapopulational recurrent selection methods may be used. However, when there is an effect of heterosis (non-additive genetics effects), interpopulation methods are more suitable to use and are able to more rapidly obtain superior genotypes for NUE.

4.6 Biotechnology Applied to Improve NUE

4.6.1 Candidates Genes and Quantitative Trait Loci

The NUE is a polygenic complex trait that is influenced by the environment, and there is considerable genetic variation in the germplasm of the main species grown (Kant et al. 2011). Consequently, several genes and metabolic pathways are associated with NUE in cereals, such as maize, rice, wheat and barley, suggesting that changes in the components of the metabolic pathway of N is not sufficient to significantly influence changes in NUE.

Quantitative trait loci (QTL) mapping provides the best understanding of the genetic control and inheritance of NUE and can be used to indicate the best selection strategy. However, functional genomics approaches are important to identify key-points in the network of genes related to N metabolism that could be modified through biotechnology. Thus, there is great effort to integrate the mapping of quantitative traits with metabolic profiles, iRNA expression profiles and transgenics to identify candidate genes that might improve NUE in maize

(Hirel et al. 2007), rice (Bi et al. 2009), wheat and barley (Sylvester-Bradley and Kindred 2009), among other important species.

Some studies report the identification of QTLs for NUE and genomic regions that might be involved with NUE in maize (Gallais and Hirel 2004), rice (Cho et al. 2007), and *Arabidopsis* (Loudet et al. 2003) under low and high N availability in the soil. Several of these QTLs are coincident with the GS and NR enzymes, which are enzymes in N metabolism that are related traits to NUE, especially under low N conditions (Hirel et al. 2007).

When QTLs of interest are identified, the molecular markers in linkage disequilibrium with the QTLs might be used (pyramiding) for genotyping because they do not occupy the same locus. The advantage of this method is that the accumulation of favorable alleles can be obtained without agronomic evaluation, saving time and labor (Fritsche-Neto 2011). However, it is important to note that there are few studies concerning the epistatic interactions between co-localized QTLs, i.e., there is no guarantee that the accumulation of QTLs of interest in a single genotype potentiate NUE.

In the future, recent advances in molecular biology and statistical genomics will impart accuracy to the identification of QTLs associated with NUE and provide consistent QTL results for NUE under different N availability in the soil to improve the phenotyping of the genotypes. Thus, large populations should be used for mapping under different locations and years to observe the maximum number of traits related to NUE. In addition, more precise experiments can be conducted in lattice using mixed models in statistical analyzes. Moreover, the characterization of the QTLs identified by sequencing analysis will allow the identification of potential structural or regulatory genes that control NUE during plant development in accordance with different N availability in the soil (Hirel et al. 2007).

4.6.2 Genome-Wide Selection

In the evaluation and selection based on the phenotype of NUE and its components, destructive methods are generally used; consequently, depending on their genetic structure, this method does not allow the identification of plants that are superior hybrids. Thus, a reduction in the potential gain is obtained with this selection. Moreover, some conventional breeding methods, such as recurrent selection, are extremely laborious and have lower gains using selection per unit time, making this selection method uneconomical.

A genome-wide selection (GWS) was proposed to increase the breeding efficiency (Meuwissen et al. 2001). With this method, the prediction and selection can be performed in juvenile plants stages, accelerating the process of breeding. Thus, GWS provides direct early selection because it is performed in genes that are expressed in the adult stage. Unlike, the early traditional selection, which is indirect because it is performed (via phenotypic evaluation) in genes activated at earlier developmental stages with the expectation that this selection would provide

Table 4.3 Estimates of the selection gains (SGs) using deviations-standard additive genetics and the relative efficiency (RE) of genome-wide selection (GWS) under one (GWS 1) or two (GWS 2) cycles per year in relation to the intrapopulation recurrent selection method (IRS) for the N uptake efficiency (NUpE), N utilization efficiency (NUtE) and shoot dry weight (SDW) under low (LN) and high N (HN) in maize

Parameter	Method	Trait			
		NUpE (HN)	NUpE (LN)	NUtE (HN)	SDW (HN)
GS	IRS	0.69	0.65	0.31	0.70
	GWS	1.80	1.32	1.67	1.70
RE (%)	GWS 1	393.84	306.98	816.99	363.82
	GWS 2	1270.80	990.53	2636.15	1173.92

Source Adapted from Fritsche-Neto (2011)

a partial understanding concerning the genes expressed in the adult stage. Thus, Fritsche-Neto (2011) compared the GWS and the intrapopulation recurrent selection method to improve the NUE components in tropical maize (Table 4.3).

According to the author, with one GWS cycle per year, the genetic gains would be 257.32 % for NUpE under low nitrogen availability and 336.03 and 603.54 % for NUpE and NUtE, respectively, under high N compared to the values obtained using intrapopulation recurrent selection. When the possibility of two GWS cycles per year is considered, these values increase to 830.27, 1084.25 and 1947.41 %, respectively. Notably, with GWS, there is an increase in selection gains per unit time and consequently, in the process of breeding. Thus, this method could revolutionize the current methods used for selection in breeding programs under abiotic stress conditions.

Although the molecular marker-assisted selection is still considered an expensive technique, these markers could reduce the time, manpower, experimental area and risks involved in developing new cultivars with high NUE.

4.6.3 Transgenes for NUE

The speed and accuracy of transgene technology allows not only the ability to test candidate genes that are critical to NUE using overexpression but also the identification of genes through mutation and gene silencing.

Studies using transgenic approaches have reported the modification of the expression of genes involved in NUE. One such modification involves the promoter of the protease gene for advanced senescence, which drives the expression of isopentenyl transferase in *Agrobacterium*. This promoter increased the levels of cytokinin, resulting in a stay green phenotype with delayed senescence and reduced photosynthetic activity and days to flowering (Robson et al. 2004). In another study, the overexpression of an isoform of the enzyme glutamine synthetase increased the grain yield in maize by approximately 30 % compared with non-transgenics (Martin et al. 2006). However, these gains were observed only in early generations of the transgenic lines grown in the greenhouse. Thus, these results must be confirmed in hybrids under low N availability in the soil.

Due to the ease of genetic transformation, a large number of genes related to NUE were tested in *Arabidopsis*, tobacco, and, more recently, rice. The genes encoding transporters for ammonium and nitrate, nitrate reductase and many key enzymes for primary N assimilation were overexpressed and, in some cases, regulated by gene silencing. These genetic transformations often led to changes in the quantity and form of N accumulation in the plant, especially under low nitrogen conditions, and in some cases, accelerated the development of the plant.

There are three notable examples of genetic transformation of plants to NUE: increase in the grain yield in rice plants overexpressing NADH-GOGAT (Tabuchi et al. 2007); higher N concentration in Arabidopsis seeds with increased asparagine synthetase activity (Lam et al. 2003); and higher grain yield under low N conditions in canola and rice overexpressing the enzyme alanine aminotransferase (Bi et al. 2009).

Another study showed an increased number of grains per panicle when the gene expression of cytokinin oxidase was reduced in the inflorescence of rice (Ashikari et al. 2005). The decreased expression of this enzyme caused the accumulation of cytokinin in the inflorescence meristem and increased the number of reproductive organs, resulting in a higher NUE.

Because NUE involves several components, there are several ways to modify gene expression and thereby increase the NUE. In addition to the plant genes and microorganisms available, other tools of genetic engineering, such as gene "shuffling" and iRNA, offer many possibilities for the improvement of NUE. Strategies to improve the absorption and assimilation of N are currently being evaluated in many plant species under field conditions and should be implemented quickly to breeding strategies. While methods for understanding the mechanisms of remobilization and utilization of N during grain development are still not defined, once discovered, they will significantly increase NUE in crop species.

References

Algren S, Baky A, Bernesson S, Nordberg A, Norén O, Hansson PA (2008) Ammonium nitrate fertilizer production based on biomass—environmental effects from a life cycle perspective. Bioresour Technol 99:8034–8041. doi:101016/S0960852408002630

Ashikari M, Sakakibara H, Yamamoto T, Takashi T, Nishimura A, Angeles ER, Qian Q, Kitano H, Matsuoka M (2005) Cytokinin oxidase regulates rice grain production. Science 309:741–745. doi:10.1126/science.1113373

Bänzinger M, Bétran FJ, Lafitte HR (1997) Efficiency of high nitrogen environment for improving maize for low-nitrogen environment. Crop Sci 37:1103–1109

Bänzinger M, Edmeades GO, Beck D, Bellon M (2000) Breeding for drought and nitrogen stress tolerance in maize: from theory to practice. CIMMYT, México

Bernardo R (2002) Breeding for quantitative traits in plants. Stemma Press, Woodbury

Bi Y, Kant S, Clark J, Gidda S, Ming F, Xu J, Rochon A, Shelp BJ, Hao L, Zhao R, Mullen RT, Zhu T, Rothestein SJ (2009) Increased nitrogen use efficiency in transgenic rice plants over-expressing a nitrogen responsive early nodulin gene identified from rice expression profiling. Plant Cell Environ 32:1749–1760. doi:10.1111/j.1365-3040.2009.02032.x

Chevalier P, Schrader LE (1977) Genotypic differences in nitrate absorption and partitioning of N among plant parts in maize. Crop Sci 17:897–901

Cho YI, Jiang WZ, Chin JH, Piao ZZ, Cho YR, McCouch SR, Koh HJ (2007) Identification of QTLs associated with physiological nitrogen use efficiency in rice. Mol Cells 23:72–79

Chun L, Mi G, Li J, Chen F, Zhang F (2005) Genetic analysis of maize root characteristics in response to low nitrogen stress. Plant Soil 276:369–382. doi:10.1007/s11104-005-5876-2

Comstock RE, Robinson HF, Harvey PH (1949) A breeding procedure designed to make maximum use of both general and specific combing ability. Agron. J. 41:360–367

DeLima RO (2010) Herança e correlação de caracteres de raiz e eficiência de uso do nitrogênio em milho. Universidade Federal de Viçosa, Viçosa, Brasil. (Thesis of Magister Scientiae)

DoVale JC (2011) Herança e correlação entre caracteres associados à eficiência do uso de nitrogênio e fósforo em milho tropical. Universidade Federal de Viçosa, Viçosa, Brasil. (Thesis of Doctor Scientiae)

Fidelis RR, Miranda GV, Santos IC, Galvão JCC (2005) Metodologias de seleção de cultivares de milho para eficiência na absorção e utilização do nitrogênio. Revista Ceres 52:987–1002

Fidelis RR, Miranda GV, Santos IC, Galvão JCC, Peluzio JM, Lima SO (2007) Fontes de germoplasma de milho para estresse de baixo nitrogênio. Pesquisa Agropecuária Tropical 3:147–153

Fritsche-Neto R (2011) Seleção genômica ampla e novos métodos de melhoramento do milho. Universidade Federal de Viçosa, Viçosa, Brasil. (Thesis of Doctor Scientiae)

Gallais A, Hirel B (2004) An approach of the genetics of nitrogen use efficiency in maize. J Exp Bot 55:295–306. doi:10.1093/jxb/erh006

Hirel B, Le Gouis J, Ney B, Gallais A (2007) The challenge of improving nitrogen use efficiency in crop plants: towards a more central role for genetic variability and quantitative genetics within integrated approaches. J Exp Bot 58:2369–2387. doi:10.1093/jxb/erm097

Kant S, Bi Y, Rothstein JS (2011) Understanding plant response to nitrogen limitation for the improvement of crop nitrogen use efficiency. J Exp Bot 62:1499–1509. doi:10.1093/jxb/erq297

Lam HM, Wong P, Chan HK, Yam KM, Che L, Chow CM, Coruzzi GM (2003) Overexpression of the ASN1 gene enhances nitrogen status in seeds of *Arabidopsis*. Plant Physiol 132: 926–935. doi:10.1104/pp.103.020123

Laperche A, Brancourt-Hulmel M, Heumez E, Gardet O, Le Gouis J (2006) Estimation of genetic parameters of a DH wheat population grown at different N stress levels characterized by probe genotypes. Theor Appl Genet 112:797–807. doi:10.1007/s00122-005-0176-z

Le Gouis J, Béghin D, Heumez E, Pluchard P (2000) Genetic differences for nitrogen uptake and nitrogen utilization efficiencies in winter wheat. Eur J Agron 12:163–173. doi:10.1016/s1161-0301(00)00045-9

Lemaire G, Charrier X, Hébert Y (1996) Nitrogen uptake capacities of maize and sorghum crops in different nitrogen and water supply conditions. Agronomie 16:231–246

Loqué D, Von Wirén N (2004) Regulatory levels for the transport of ammonium in plant roots: regulatory aspects of nitrogen assimilation. J Exp Bot 55:1293–1305. doi:10.1093/jxb/erh147

Loudet O, Chaillon S, Merigout P, Talbotec J, Daniel-Vedele F (2003) Quantitative trait loci analysis of nitrogen use efficiency in *Arabidopsis*. Plant Physiol 131:345–358. doi:10.1104/pp.102.010785

Machado AT (1997) Perspectiva do melhoramento genético em milho (*Zea mays* L.) visando eficiência na utilização do nitrogênio. Universidade Federal Rural do Rio de Janeiro, Rio de Janeiro, Brasil. (Thesis of Doctor Scientiae)

Machado AT, Magalhães JR, Magnavaca R, Silva MR (1992) Determinação da atividade de enzimas envolvidas no metabolismo do N em diferentes genótipos de milho. Revista Brasileira de Fisiologia Vegetal 4:45–47

Martin A, Lee J, Kichey T, Gerentes D, Zivy M, Tatout C, Dubois F, Balliau T, Valot B, Davanture M, Tercé-Laforgue T, Quilleré I, Coque M, Gallais A, Gonzalez-Moro MB, Bethencourt L, Habash DZ, Lea PJ, Charcosset A, Perez P, Murigneux A, Sakakibara H, Edwards KL, Hirel B (2006) Two cytosolic glutamine synthetase isoforms of maize are

specificafilly involved in the control of grain production. Plant Cell 18:3252–3274. doi:10.1105/tpc.106.042689
Meuwissen T, Hayes B, Goddard M (2001) Prediction of total genetic value using genome-wide dense marker maps. Genetics 157:1819–1829
Miranda GV, Godoy CL, Souza LV, Santos IC (2005) Selection of discrepant maize genotypes for nitrogen use efficiency by a chlorophy meter. Crop Breed Appl Biotechnol 5:451–459
Moll RH, Kamprath EJ, Jackson WA (1982) Analysis and interpretation of factors which contribute to efficiency of nitrogen utilization. Agronomy J 74:562–564
Oliveira LR (2009) Eficiência de uso de nitrogênio e atividade da nitrato redutase e glutamina sintetase em milho. Universidade Federal de Viçosa, Viçosa, Brasil. (Thesis of Doctor Scientiae)
Presterl T, Seitz G, Landbeck M, Thiemt W, Schmidt W, Geiger HH (2003) Improving nitrogen use efficiency in European maize: estimation of quantitative parameters. Crop Sci 43:1259–1265
Roach DA, Wulff RD (1987) Maternal effects in plants. Annu Rev Ecol Syst 18:209–235
Robson PR, Donnison IS, Wang K, Frame B, Pegg SE, Thomas A, Thomas H (2004) Leaf senescence is delayed in maize expressing the *Agrobacterium IPT* gene under the control of a novel maize senescence enzymes of maize hybrids representing seven eras of maize breeding. Maydica 43:83–94
Souza LV, Miranda GV, Galvão JCC, Eckert FR, Mantovani EE, Lima RO, Guimarães LJM (2008) Genetic control of grain yield and nitrogen use efficiency in tropical maize. Pesquisa Agropecuária Brasileira 43:1517–1523. doi:10.1590/S0100-204X2008001100010
Souza LV, Miranda GV, Galvão JCC, Guimarães LJM, Santos IC (2009) Combining ability of maize grain yield under different levels of environmental stress. Pesquisa Agropecuária Brasileira 44:1297–1303. doi:10.1590/S0100-204X2009001000013
Sylvester-Bradley R, Kindred DR (2009) Analysing nitrogen responses of cereals to prioritize routes to the improvement of nitrogen use efficiency. J Exp Bot 60:1939–1951. doi:10.1093/jxb/erp116
Tabuchi M, Abiko T, Yamaya T (2007) Assimilation of ammonium ions and reutilization of nitrogen in rice (*Oryza sativa* L.). J Exp Bot 58:2319–2327. doi:10.1093/jxb/erm016
Wissuwa M, Mazzola M, Picard C (2009) Novel approaches in plant breeding for rhizosphere-related traits. Plant Soil 2:409–443. doi:10.1007/s11104-008-9693-2

Chapter 5
Breeding for Phosphorus Use Efficiency

Sidney Netto Parentoni, Flávia Ferreira Mendes and Lauro José Moreira Guimarães

Abstract Low phosphorus availability is one of the major agricultural limitations in tropical soils. Phosphorus deficiency causes a series of effects in plants such as changes in root and shoot morphology, in plant physiology and in plant internal transport. Phosphorus deficiency also affects the use of other nutrients, resulting in reduction of crop yield potential. Among the macronutrients, phosphorus has the lowest use efficiency in plants. This low efficiency is related to the ability of soils to adsorb the applied P, making it unavailable to the crop. Phosphorus efficiency has been defined as the processes by which plants acquire, translocate, accumulate, and utilize this nutrient to better produce dry matter and/or grain under conditions of high and low supply. Nutrient use efficiency has been defined as grams of grain produced per gram of nutrient supplied to the crop. This index can be decomposed into two main components: acquisition efficiency, and internal utilization efficiency. Phosphorus acquisition efficiency has been found to be two to three times more important than phosphorus internal utilization efficiency in tropical soils. Breeding programs for phosphorus use efficiency should take into account a number of issues such as: use of appropriate screening sites, germplasm sources, experimental conditions, intensity of stress imposition, selection criteria, genetic information and evaluation methods. A general discussion of these topics are covered in this chapter.

Keywords Phosphorus · P use efficiency · Breeding · Maize · Nutrient · Stress

S. N. Parentoni (✉) · L. J. M. Guimarães
EMBRAPA-Maize and Sorghum Research Center, Sete Lagoas, Brazil
e-mail: sidney@cnpms.embrapa.br

L. J. M. Guimarães
e-mail: lauro@cnpms.embrapa.br

F. F. Mendes
UFLA—Federal University of Lavras, Lavras, Brazil
e-mail: flvmendes2001@yahoo.com.br

R. Fritsche-Neto and A. Borém (eds.), *Plant Breeding for Abiotic Stress Tolerance*,
DOI: 10.1007/978-3-642-30553-5_5,

5.1 Introduction

Significant gains in productivity and production of different crops have occurred throughout the world during the last two decades, and part of these gains have been obtained by an increase in tolerance to biotic (pests and diseases) and abiotic stresses (toxicity/mineral deficiencies, thermal variations, and drought). The FAO report (2010) predicts an expected population growth close to 34 % by 2050, with the population of the planet reaching 9.1 billion people. To meet the anticipated demand for food, the world production is expected to grow 70 % by the mid-century. Accordingly, in addition to increase productivity, new areas will have to be incorporated into agricultural production, especially in the American and African continents. It is estimated that close to 120 million ha should be integrated into production systems in these regions and land previously occupied by livestock farming with low use of inputs should become cultivated farmland. In general, the soils in these regions are highly weathered, have a moderate-to-high degree of acidity, toxic levels of Al and low availability of nutrients, especially phosphorus (P).

Norman Borlaug, father of the "Green Revolution" and winner of the Nobel Peace Prize in 1970, said that, with the population growth expected in the coming decades, genetic improvement of the major crops will play a key role to meet the large demand for food and collective efforts of conventional breeding and Biotechnology will be necessary to overcome yield losses due to abiotic stress (Borlaug 2000). To increase production, a better understanding of agricultural practices such as liming, fertilization, irrigation, pest, disease and weed control, direct planting, crop rotation, among others, will also play a relevant role.

In tropical soils, low P availability has been considered as one of the major agricultural limitations. Phosphorus is one of the six macronutrients essential for plant growth. In addition to being a structural element of nucleic acids and phospholipids, it also participates in the processes of energy production and enzyme activation. Phosphorus deficiency causes a series of effects in the plant such as changes in root and shoots morphology, changes in plant physiology and in plant internal transport. It also affects the use of other nutrients, resulting in reduction of yield potential of crops.

The phosphorus (P) supply in tropical soils has been accomplished by the use of fertilizers. The amount of P fertilizer required during the crop cycle is high. Baligar et al. (2001) compared the average efficiency of phosphorus used by plants. The authors reported that in general, P use efficiency in tropical soils can be lower than 10 %. This low efficiency of P use is related to the ability of soils to adsorb the applied P making it unavailable to the crop and also to the plant mechanisms related to both acquisition and internal utilization efficiency of this nutrient by the plants.

The search for cultivars that could utilize P more efficiently has three basic reasons: (i) fertilizers are inputs with greater participation in agricultural production costs; (ii) the indiscriminate use of fertilizers in high doses has large environmental impact and may contaminate water sources; (iii) the production of phosphate fertilizers depends on non-renewable mineral sources. Estimates of time-line of

Table 5.1 Estimates of required time to consume the phosphate reserves that can be exploited at a cost of less than $40.00 or less than $100.00 per ton

Country	Time necessary for reserves consumption (years)	
	<US$40 ton^{-1}	<US$100 ton^{-1}
	number of years	
Brazil	75	84
China	46	457
Morocco	248	913
Russia	20	100
Tunisia	13	76
EUA	25	98
World	90	343

Source adapted from Murrel and Fixen (2006)

P reserves point out that worldwide reserves with exploitation costs of less than 40 dollars/ton will be exhausted in this century (Murrel and Fixen 2006), as shown in Table 5.1. Accordingly, the development of plants with higher P utilization efficiency should become increasingly important.

5.2 Phosphorus Dynamics in the Soil

Phosphorus does not occur as a single element in nature. It is always combined with one or more elements, producing various forms of phosphate with different degrees of complexity. In this text, the word phosphorus or the symbol P will normally be used without referring to a particular type of phosphate.

Soil phosphorus is divided into two major groups, the inorganic and the organic one. In most cases, the concentration of inorganic phosphorus in the soil solution is extremely low, ranging from 0.1 to 10 μM, which means that the plants must absorb this nutrient from extremely dilute solutions (Loughman et al. 1983).

In addition to the low concentration of this nutrient, tropical soils have a high capacity of fixing it, mainly in amorphous hydrated forms of iron oxide (hematite), aluminum oxide (gibbsite), and kaolinite-type clay. This type of fixation is related to the process of soil formation. Phosphorus tends to be precipitated as aluminum phosphates in more humid environments and as iron phosphates in drier environments (Novais and Smyth 1999).

The absorption of inorganic phosphorus also depends on the pH of the soil. In soils with a pH below 7, the concentration of the $H_2PO_4^-$ anion is higher. This is the form preferentially absorbed by plants. However, in acid soils, the concentrations of Fe and Al in the solution also increase, leading to the precipitation of inorganic phosphorus as iron and aluminum phosphates. With an increase in pH, a lower degree of adsorption of phosphate on the adsorbent surface occurs. However, at very alkaline pH, almost all inorganic P is in the PO_4^{3-} form, and this

Fig. 5.1 The effect of the pH on the availability of soil P for plants. Adapted from Hyland et al. (2005)

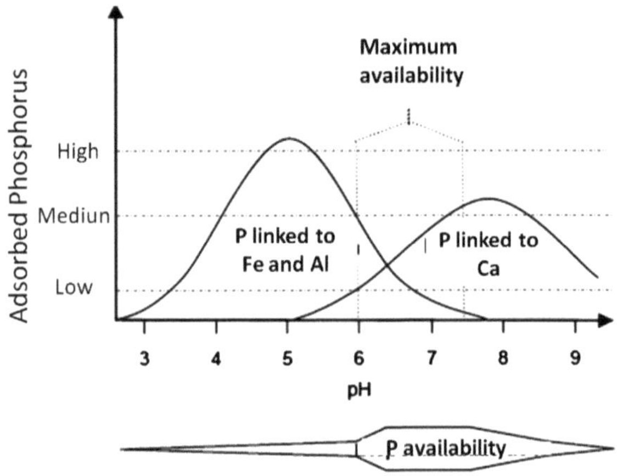

form is not absorbed by plants. In addition in calcareous soils with a pH above 7, P can be precipitated as hydroxyapatite or carbonate apatite (Dechen and Nachtigall 2007). The soil–plant system is dynamic, and it is difficult to define the optimum pH for the various annual crops. In general, the maximum availability of phosphorus to plants occurs at pH values between 6 and 7 (Fig. 5.1).

Organic P can constitute 5–80 % of the total P in the soil. The more common sources are the inositol phosphates, comprising 10–80 % of the total organic phosphorus, phospholipids (0.5–7 %), nucleic acids (~3 %), and other esters phosphate (>5 %) (Dalal 1977). Another fraction that can contain P in the soil is the microbial biomass. During the growing season, the mineralization of organic matter results in a slow release of P into the soil solution, making it available for absorption by the plant roots.

Hisinger et al. (2006) found that the volume of soil explored by the roots is only 0.4–0.8 % of the total volume of the surface layer, dramatically limiting the absorption of P, which occurs by the diffusion mechanism (and not through mass flow, as seen in the case of other nutrients such as N). This diffusion requires a close proximity between the root and the nutrient in the soil, so that the size and area of the root system are key points in determining the ability of plants to acquire P from the soil. Also, the diffusion process only occurs in the presence of water, which can be a common point between P and drought stresses.

Practices such as liming and the increase in content of organic matter tend to increase P availability in tropical soils. In soils under no-till cultivation in Brazil, a large fraction of phosphorus is found in the organic fraction compared to the inorganic fraction and this proportion is largest in no-till areas than in conventional soil preparation areas (Anghinoni 2006). From the standpoint of phosphorus nutrition, it is clear that the maintenance of soil organic matter at appropriate levels is an important factor for better crop development.

5.3 Concepts of P Use Efficiency

To understand the process of evaluating the efficiency of the use of phosphorus by plants, a clear definition of the concepts of efficiency is necessary. P efficiency has been defined as the processes by which plants acquire, translocate, accumulate, and utilize the nutrient to better produce dry matter and/or grains under conditions of high and low supply of the nutrient (Cicarelli et al. 1998). Moll et al. (1982) defined the efficiency of use of a nutrient as the number of grams of grain produced per gram of nutrient supplied to the crop. This index can be decomposed into two main components. The first is called the "efficiency of acquisition", defined as the amount of nutrient uptake by plants divided by the quantity of nutrients available to the crop in the soil. The second component is called "internal utilization efficiency", defined as the weight of grain produced per unit of nutrient uptake by the plants. The internal utilization efficiency can also be divided into two components: the plant's capacity to take the nutrient from the shoot to the grain (harvest index of the nutrient) and the plant ability to produce grain per unit of nutrient in the grain (nutrient utilization quotient).

The P use efficiency (E_{USE}) and its components can be obtained from the following equations:

$$E_{USE} = E_{AQ} \times E_{UTIL}$$

where

E_{USE}: P use efficiency;
E_{AQ}: P acquisition efficiency;
E_{UTIL}: P internal utilization efficiency.

The P acquisition efficiency is obtained from:

$$E_{AQ} = (P_{Tot}/P_S)$$

where

P_{Tot}: total P content in the upper part of the plant (stem + leafs + grain) and
P_S: quantity of available P in the soil.

The P internal utilization efficiency is obtained from:

$$E_{UTIL} = (DMG/P_{Tot})$$

where

DMG: dry mass of the grains.

The unit of measure used to express the amount of P in the soil and in the plant must be the same because the efficiency ratios return dimensionless values.

Moll et al. (1982) proposed a method to determine the relative importance of the two components in the efficiency of use of a particular nutrient. They used a logarithmic transformation to obtain an additive relationship between these variables ($\log_{EUSE} = \mathrm{Log}_{EAQ} + \mathrm{Log}_{EUTIL}$). The transformation of the components must be made for each genotype using the observations in a plot basis. Once this relationship is obtained, it can be determined the relative importance of the sum of squares of E_{AQ} and E_{UTIL} in the sum of squares of E_{USE} (which will determine the relative importance of E_{AQ} and E_{UTIL} to the E_{USE}).

Using this method, Parentoni and Souza Junior (2008) determined the relative importance of the efficiency of acquisition and internal utilization efficiency in phosphorus use efficiency of maize grown in tropical savanna soils. The authors found that P acquisition efficiency was almost two times more important than P internal utilization efficiency in environments with low P availability, and E_{AQ} was almost three times more important than E_{UTIL} at high P environments. Similar results were reported for wheat, soybeans, and rice (Wang et al. 2010). The lack of correlation between the efficiency of acquisition and internal utilization efficiency for P was reported by Parentoni and Souza Junior (2008). This result indicates that selection in one of these variables would not affect the other in a predictable way. Accordingly, breeding programs with the goal to produce cultivars which are more efficient in P use should apply selection indexes that give greater weights for P acquisition efficient than for P internal utilization efficiency.

5.4 Genetics of Phosphorus Efficiency

Genetic variability for phosphorus use efficiency has been reported in a number of crops such as rice, wheat, bean, soybean, and maize (Parentoni et al. 2010; Zhang et al. 2009; Ochoa et al. 2006; Cicarelli et al. 1998; Barriga and Proschle 1996 and Chaubey et al. 1994). Phenotypic variation in maize inbred lines with high and low P use efficiency grown in a low P soil is shown in Fig. 5.2.

For the choice of suitable breeding methods for a given trait, it is important to know the genetic control and the predominant type of gene action for the characteristic of interest. Efforts have been made to understand and explore the variability of the characteristics related to the P use efficiency by plants, which has proved to be a complex task because of the fact that the inheritance of these characteristics is generally polygenic and highly influenced by the environment.

A large portion of the studies in the genetic control of the P use efficiency is based on evaluation of acquisition and accumulation of phosphorus at the seedling stage, in sorghum, wheat, and maize (Gorz et al. 1987; Ahsan et al. 1996; Furlani et al. 1998). These studies have shown that both additive and non-additive effects are important in the control of this trait. However, it must be emphasized that results obtained in nutrient solution, or even in pots, often do not correlate well with grain yield in field studies. In this case, experiments conducted under field conditions are needed for a better understanding of the genetic of traits related to P use efficiency.

Field-level results were obtained by Parentoni et al. (2000). The authors evaluated a maize inbred lines diallel under high and low P soil. It was found that non-additive genetic effects were larger than additive effects for the trait grain weight produced per unit of P in the grain (quotient of utilization). Similar results were found by Parentoni (2008) for P acquisition efficiency and P internal utilization efficiency in maize.

Fig. 5.2 Plots of two maize inbred lines grown in a low P soil (*left* picture) and in a high P soil (*right* picture). The maize inbred line L3 is highly P use efficient and the maize line L53 is an inefficient line. Photos taken by Sidney Parentoni

Chaubey et al. (1994) studied the genetics of P use efficiency in rice (measured as the relative ability to emit tilers under contrasting P conditions). The authors concluded that there was a predominance of non-additive genetic effects in the control of this trait in rice. Barriga and Proschle (1996) evaluated a diallel among wheat lines for the traits P content in shoots, P harvest index (grain P/total P in the plant) and P internal utilization efficiency (grain dry matter/total P in the plant). These authors found a predominance of non-additive genetic effects and high heterosis for all traits studied.

5.5 Mechanisms Related to P Use Efficiency in Plants

The cultivation of plants in soils with low P levels has resulted in the development of a variety of strategies for increasing the acquisition and or improving the use efficiency of this nutrient. The main mechanisms associated with adaptation to P stress in plants can be divided into two groups: mechanisms related to the nutrient acquisition efficiency and those related to P internal utilization efficiency (Fig. 5.3).

The major mechanisms related to P acquisition efficiency include:

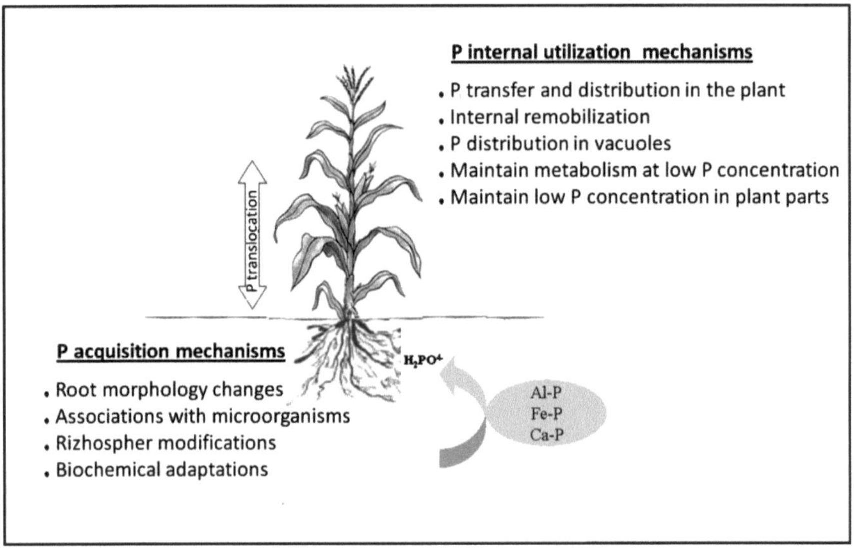

Fig. 5.3 Main mechanisms of P acquisition and P internal utilization efficiency in plants

1. *Root modifications*—increase in the root volume and in the amount of root hairs (related to a large soil volume exploitation); increase in the release of organic acids, formation of aerenchyma in root tissue (which reduces the respiration rate of roots and, consequently, the P requirement in these tissues).

2. *Associations with microorganisms*—related to the plant's capacity to form symbiotic associations with mycorrhiza and phosphate-solubilizing bacteria. The associations with mocorrízic fungi increase P availability to plants as a result of the increase in effective surface area for absorption by the roots in the soil via the association between mycorrhizal hyphae and roots. The release of exudates (such as organic acids) by plant roots can alter the microflora in the rhizosphere, leading to an increase in the population of P-solubilizing microorganisms in this region.

3. *Modifications in the microenvironment of the rhizosphere*—some plants are able to secrete organic acids and phosphatases, which act in the dissolution of P.

4. *Biochemical modifications*—changes in the membrane transporters related to the absorption of P.

The mechanisms related to the internal utilization efficiency can be divided into the following components:

1. *Transport*—transfer of the nutrient in the endoderm, release of it in the xylem, control of the rate of absorption, and distribution of nutrients in the plant.

2. *Remobilization*—translocation of P from the plant mature parts, such as senescent leaves, to the vegetative and reproductive growth meristems.

3. *Distribution*—directing the nutrient to cellular compartments such as vacuoles.

4. *Metabolism*—the genotype's capacity to maintain the normal metabolism with low concentrations of nutrient in the tissue.

5. *Allocation*—maintaining low concentrations of the nutrient in the physical support structures such as stalks or stems.

An additional variable that has been evaluated in maize under stress conditions is the interval between the male and female flowering. The lack of synchrony between them can lead to lower grain yield because scarcity of pollen available to fertilize the style-stigma and also because kernel abortion shortly after fertilization. An increase in the interval between male and female flowering (ASI) in environments that causes stress in plants have been extensively reported under drought conditions (Bolanos and Edmeades 1993). Parentoni et al. (2010) found that P stress also increase ASI in maize. The authors suggested that the evaluation of ASI under P stress conditions could become an auxiliary selection criterion for maize adaptation to low P stress environments.

5.6 Imposition and Intensity of Stress

Early screening for P use efficiency in maize has been suggested in a number of breeding programs. In general, these trials are performed in nutrient solution or green-house pots with restricted availability of P in the substrate. In this case, the plants are evaluated at the vegetative phase for characteristics related to its development, to dry mass accumulation and to P content in both shoots and roots.

A low correlation between data obtained by methods of early selection with the development of plants evaluated under field conditions has been observed. Although expensive, the most reliable method to determine the P use efficiency in a crop is the evaluation of genotypes in the field using contrasting areas with respect to P availability and conducting the experiments until maturity. Thus, the experimental areas for these trials should be well characterized, and the stress level must be effective for selecting superior genotypes.

5.7 Soil Characterization of Experimental Sites

The characterization of experimental sites is an important step in the evaluation of genotypes for their efficiency of P utilization. The procedures for the characterization of soils to establish specific experimental sites for cereal and leguminous crops are described by Viana et al. (2007). Initially, the site for phenotyping stress must be as representative as possible of the agricultural areas for which the genetic material is intended.

According to these authors, the main step in the characterization of these sites includes the choice of areas based on soil surveys and other information, such as aerial photographs, scientific papers and reports already available. In this sense, the history of the areas, including data from tests or activities conducted in these areas since the beginning of their occupation, are important documents.

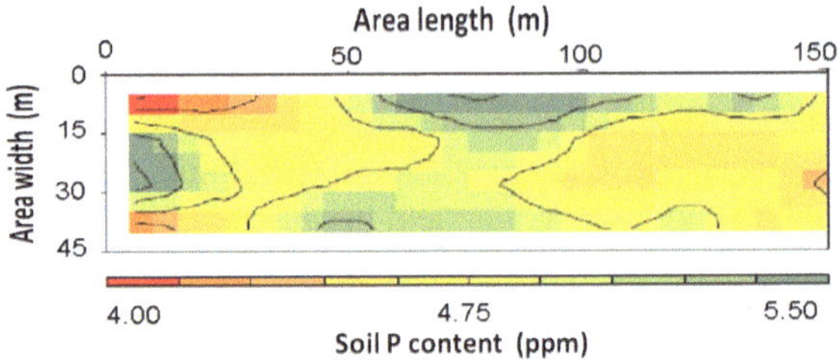

Fig. 5.4 Map of the interpolation of P grid analysis in a screening site at EMBRAPA-Maize and Sorghum. *Source* Viana et al. (2009)

After choosing the area, maps must be prepared or preliminary images obtained to aid in identifying different soil and vegetation patterns. A planialtimetric survey should be performed, and points of reference should be registered for use in data collection. A detailed study of soils, with the opening of a trench, the description of profiles and a survey of properties such as texture, fertility, resistance, infiltration curves, thickness of the A horizon, textural gradients, and electrical conductivity is recommended. The establishment of small grids for collecting composite soil samples at depths of 0–10, 10–30 and 30–50 cm in each grid, with georeferencing of the sampling points is suggested. Finally, the compilation of these data in a geographic information system and generation of thematic maps and numerical models of the area should be performed.

A map showing the interpolation of the results obtained for a P screening site at EMBRAPA-Maize and Sorghum Research Center is shown in Fig. 5.4. An area of 6,776 m^2 (150×45 m) was divided into small grids of 11×14 m each, where soil samples were collected.

According to the map of interpolation analyses, the P concentration in the soil ranged from 4.00 to 5.50 ppm (Fig. 5.4). The color code indicating P concentration facilitates the identification of possible fertility spots in the soil and enables the allocation of experimental blocks into homogeneous areas. With more detailed information about the phenotyping sites, better adjustments may be made to standardize the conditions for the evaluation of the genotypes of interest.

5.8 Intensity of the Stress

In addition to the site chemical and physical characterization, it is important that the stress induction in the trials would be effective. The stress should be able to reduce yield potential by 40–60 % relative to the values obtained in non-stress environment. Experimental data have shown that, if the environment with a low P level causes a small reduction in the yield potential, the genetic correlation

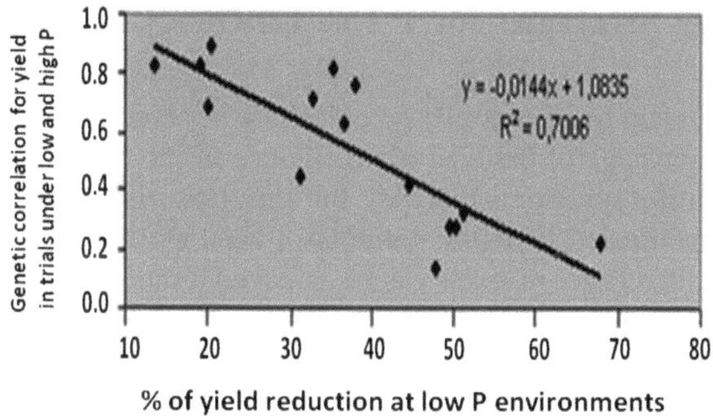

Fig. 5.5 Relationship between genetic correlation coefficient and percentage of yield reduction for 15 different maize pairwise trials under low and high P (% of yield reduction was calculated as grain yield at high P minus grain yield at low P divided by grain yield at high P). *Source* adapted from Parentoni et al. (2010)

between the materials evaluated in the presence and absence of P stress tend to be relatively high (Parentoni et al. 2010), as can be seen in Fig. 5.5. However, it is expected that a very high stress intensity will lead to a large genotype x environment interaction which would make more difficult to select for both environments. In this case, the classification of the genotypes will be altered when they are evaluated in contrasting environments. The level of stress also, should not drastically reduce the genetic variability as a result of loss of plants in the plots or by restricting productivity, with values close to zero, because the identification of superior genotypes would be compromised in this situation.

For example, experiments with maize conducted under conditions of P deficiency in the region of Sete Lagoas, MG have been allocated in areas with an availability of this nutrient ranging from 2 to 6 ppm. In this region, dystrophic red oxisols predominate, and the control experiments, i.e., those experiments without stress resulting from P deficiency have been allocated to areas with P levels of 12–15 ppm and with the application of approximately 100 kg of P_2O_5 per hectare. Under these conditions, a decrease in grain yield close to 50 % in the environments with P stress compared to that obtained in non-stress environments has been considered adequate for selection for P efficiency.

It is worth mentioning that the correction of soil acidity is a prerequisite both for areas with and for areas without stress resulting from P deficiency so that the effects caused by the acidity can be separated from those related to the P use efficiency. Nitrogen fertilization at planting and side-dressed, as well as the application of potassium, should be the same in both P level trials. It should also ensure the availability of water during the crop cycle to prevent drought stress so that drought tolerance will not be confused with nutrient use efficiency.

5.9 Breeding Strategies for P Use Efficiency in Maize

A breeding program to increase P use efficiency in plants can be planned on the basis of the knowledge of the type of inheritance of the trait of interest. The first step is to identify superior genotypes for this trait (inbred lines, commercial hybrids, open pollinated varieties, germplasm bank entries, etc.). One must consider, however, that faster progress can be achieved if the initial breeding materials would contain genotypes already used in the target region and with the greatest degree of breeding as possible. Otherwise, a number of problems could arise from germplasm lack of local adaptation and due to the presence of a high frequency of deleterious alleles in the materials.

As discussed before, both additive and non-additive effects are important in the genetic control of P use efficiency. However, non-additive genetic effects predominate in several agricultural crops such as maize, sorghum, wheat, rice, and beans (Parentoni 2008).

For autogamous plants, where inbred lines are used commercially, the formation of breeding populations should be based on crosses between genotypes that present high frequencies of favorable alleles for P use efficiency (or its components). In this case, classical methods of autogamous crop breeding may be used in the conduction of segregating populations and obtaining superior genotypes. The genealogical methods (pedigree), the bulk segregating (population method), and the single seed ascendant (SSD) method and its variations are the most common ones. Methods such as recurrent selection and the use of double haploids can also be applied.

For crops where the main commercial products are hybrids, the exploitation of heterosis via hybridization is the best strategy. Methods of interpopulacional selection can also be used. It is noteworthy that, in addition to the typical allogamous plants such as maize, breeders of autogamous plants have successfully exploited heterosis in crops such as sorghum and rice.

5.10 Experimental Conditions

Data quality is a key factor in identifying superior genotypes. Thus, some statistical concepts are useful in the experimental design to increase the test precision under stress conditions. The coefficient of variation, for example, is normally used as a measure of experimental precision. It has been observed that experiments conducted under severe stress conditions usually have high coefficients of variation. This is usually a result of the association of low means obtained in these experiments with medium-to high-magnitude environmental variances relative to those observed in trials conducted in non-stress environments.

Another parameter that can be used as a measure of precision is heritability. This estimate is more appropriate because, in addition to the residual variance, it takes into account the genetic variance and the number of replications. Furthermore, heritability it is not influenced by the mean, as it is the case with the coefficient of variation

The use of a larger number of replications is a strategy to obtain higher estimates of heritability and, consequently, to improve experimental precision under stress conditions. However, the use of large numbers of replications leads to practical limitations and it is expensive for agricultural experimentation.

Incomplete blocks, such as lattices (square, rectangular and alpha lattices) are types of experimental designs that can contribute to increase experimental precision. In large field experiments, this approach is more advantageous than the use of a randomized complete block designs because it enables a better control of environmental heterogeneity.

One technique that has been increasingly adopted in breeding programs is the use of mixed linear models. In animal breeding programs and in forest species, this method has allowed the analysis of experiments with high degree of unbalance. Thus, instead of obtaining phenotypic means, genotypic values can be predicted using this technique (de Resende 2007). The use of spatial analysis is another way of treating experimental data which can improve experimental precision, particularly when a clear trend is observed for inequality in the development of plants in the field. These trends can be caused, for example, by soil spots.

Despite the aforementioned possibilities, the most efficient way to obtain reliable estimates for selection is a careful conduction of the experiments. This procedure includes: selection of areas as homogeneously as possible; use of appropriate experimental designs; use of a larger number of replications in the experiments conducted under stress conditions compared to the ones under non stress; a detailed collection of data involving measurements, careful note taking and digitalization; implementation of appropriate cultivation practices (soil preparation, planting, weed, pest, and disease control, additional fertilization, irrigation and others). To guarantee the plant number in the plots, it is useful to plant twice more seeds as the desired plant number in a plot and to do a thinning operation after seedlings have been established.

5.11 Methods of Evaluation of P Use Efficiency in Plants

The characterization of the P use efficiency typically has been accomplished by the evaluation of experiments conducted under conditions of low availability of this nutrient. However, the concepts of efficiency and response to a given nutrient should be considered together in the identification of superior genotypes in a breeding program.

Batten et al. (1984) defined, in a practical manner, efficient and responsive cultivars with respect to the use of nutrients according to two situations: (i) efficient cultivars would be the ones able to produce higher yield with a lower dose of the fertilizer than that commonly used for the crop (deficiency condition); and

(ii) responsive cultivars would be the ones able to achieve economically significant increases in productivity with quantities of fertilizer higher than that normally used in a given environment (optimized conditions).

A technique to group genotypes according to their efficiencies and responsiveness to a given nutrient was presented by Fageria and Baligar (1993). These authors used scatter plots, in which the abscissa corresponded to the efficiency and the nutrient response was projected on the y axis. Yield under a stress condition was used as an index of efficiency in the use of P, while the response can be obtained from the expression [(GYH−GYL)/(H−L)], where GYH and GYL correspond to grain yield in environments with high and low nutrient availability, respectively, and (H−L) is the difference in nutrient availability between the levels of P (Furtini 2008).

The performance of the genotype under conditions of high nutrient availability can also be considered as a response. This approach can be advantageous when compared to the use of response indexes because it makes possible a direct identification of genotypes with superior performance in environments with and without stress. In this case, the performance in an environment without stress is not masked by indexes of ratios of productivity to nutrient availability. This methodology allows the classification of the cultivars into four groups, depending on the yield in contrasting environments with respect to the availability of P:

1. Efficient and non-responsive cultivars (ENR) are those with a high productivity under conditions of low P availability and which do not respond to an increased supply of the nutrient. They present a below-average production in the environment without stress.
2. Efficient and responsive cultivars (ER) present an above-average production both in the environment with low P availability and in the environment with high P availability. From the viewpoint of plant breeding, this situation would be ideal for selecting superior genotypes.
3. Inefficient and responsive cultivars (IR) are those with below-average yield under stress conditions as a result of P deficiency, but responding positively to an increased nutrient supply, with above-average production in an environment without stress.
4. Inefficient and non-responsive cultivars (INR) are those with below-average yields in both environments.

The productivity observed under stress conditions can be represented on the abscissa (x) of a graph and the productivity in a stress-free environment can be plotted on the ordinate axis (*12*). Genotypes in each of the four categories can be identified by two straight lines that intersect at the means of the environments. A scatter plot of the yields obtained in contrasting environments with respect to the availability of P and the identification of efficient and responsive genotypes is illustrated in Fig. 5.6.

The data shown in the above chart was obtained from maize experiments conducted at low and high P levels in the soil at EMBRAPA-Maize and Sorghum Research Center in Sete Lagoas, Brazil. A mean reduction in grain yield of 68 % from high to low P soils was observed. The correlation between the average performance of genotypes in environments with and without stress was 0.17,

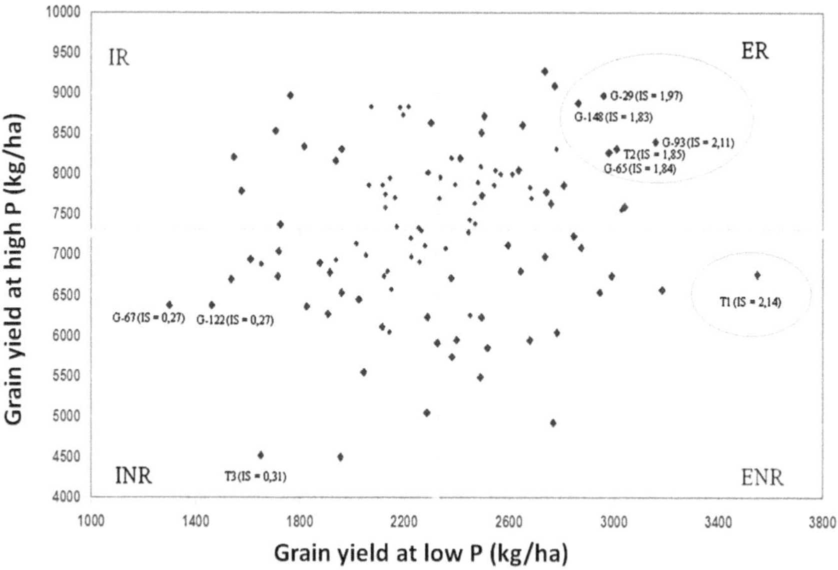

Fig. 5.6 Mean grain yield for maize genotypes cultivated at low and high P levels in the soil. *Source* Mendes (2011, unpublished data)

indicating that there was a strong genotype versus environment interaction (GE). This interaction confirmed the need to evaluate genotypes in contrasting P conditions to identify more efficient and responsive genotypes. However, even with a strong GE interaction, it is possible to select individuals with a superior performance in both conditions.

Selection indexes exist that can assist in the selection of more efficient and responsive genotypes. Parentoni et al. (2001) presented several of these indexes, including a selection index (SI), which simultaneously considers the yield at the two levels of available P. In this index, a weight is used to permit the selection of genotypes with a trend to be more efficient than responsive in P use. This selection index is presented below.

$$SI = [(Y_{iS} \times Y_{iNS})/(M_S \times M_{NS})] \times (Y_{iS}/M_S)$$

where

Y_{iS} = mean yield of genotype i in the stress environment
Y_{iNS} = mean yield of genotype i in the non-stress environment
M_S = overall trial yield mean in the stress environment
M_{NS} = overall trial yield mean in the non-stress environment

The indexes obtained using the above expression for the same 110 genotypes plotted in Fig. 5.6 are summarized in Table 5.2. The indexes ranged from 0.27 to 2.14. Considering a selection intensity of 5 % based on the index, six superior genotypes were identified, from which five were considered to be efficient and responsive, and one was the most efficient genotype, although an unresponsive

Table 5.2 Selection Index (SI) for a subset of 110 maize genotypes evaluated for grain yield (kg ha^{-1}) in environments with low P (GY_S) and with high P (GY_{NS}). Sete Lagoas, MG, 2010

Classification (IS)	Genotype	GY_S	GY_{NS}	Index (SI)
1	T-1	3,549	6,762	2.14
2	G-93	3,158	8,402	2.11
3	G-29	2,958	8,972	1.97
4	T-2	2,985	8,274	1.85
5	G-65	2,975	8,269	1.84
6	G-148	2,863	8,880	1.83
108	G-122	1,460	6,376	0.34
109	T-3	1,649	4,519	0.31
110	G-67	1,299	6,373	0.27
Mean		2,332	7,316	

Source Mendes (2011, Unpublished data)

one. However, this genotype presented a yield performance close to the mean obtained in the environment with high P availability. This selection can also be seen in Fig. 5.6, in which the selected genotypes were shown within a circle. The advantage of using an index is that data can be compared from different experiments with large variation in the trials means.

5.12 Advances in Biotechnology for P Use Efficiency

Most studies of the characteristics related to P use efficiency have demonstrated the existence of a complex inheritance pattern in several species. The mapping of chromosomal regions controlling quantitative trait loci or QTLs is a molecular strategy that can help in understanding the genetics and the mechanisms involved in the control of complex traits. It may also offer possibilities for use in breeding programs via molecular marker-assisted selection, genomic selection or transgenic approaches.

Numerous QTL for P efficiency have been identified in crops such as beans (Ochoa et al. 2006), maize (Chen et al. 2009), soybean (Zhang et al. 2009), and wheat (SU et al. 2009). Most of these QTL are associated with larger root development and the capacity of plants to release organic acids that can dissolve P in the soil.

A QTL with a large effect on P acquisition efficiency in rice was identified in two independent studies using distinct methods and populations (NI et al. 1998; Wissuwa et al. 1998). The QTL, denominated *Pup 1,* (P Uptake 1) was identified on rice chromosome 12 and explained up to 80 % of the phenotype variation observed in P acquisition efficiency in a rice population evaluated under P deficiency (Wissuwa et al. 2002). Efforts have been made toward the identification of potential candidate genes in this QTL region. An ongoing research project funded by the Generation Challenge Program (GCP), which involved researchers from the University of Maiseell, IRRI (International Rice Research Institute), the JIRCA in Japan and EMBRAPA-Maize and Sorghum, sought to clone the gene responsible for this QTL and searched for the homologs in maize and sorghum. The validation of the candidate gene for QTL *Pup 1* is expected to be completed in 2012.

Although *Pup1* confers a high P use efficiency in the rice line Kasalath in which this QTL has been identified, this line cannot be recommended for commercial use because it contains several undesirable agronomic traits (Wissuwa et al. 2002). This QTL must, therefore, be transferred to genotypes with better agronomic performance. This transfer can be accomplished using the method of backcrossing with Kasalath as donor parent of the favorable QTL allele, followed by marker-assisted selection to facilitate the process.

With the advances made in molecular biology, new QTLs and/or genes are been identified and related to the increased efficiency of P use. Studies have been conducted to improve P acquisition by means of genetic engineering. The focus of some of these studies has been to look at genes that act in the production and release of organic compounds that increase P solubility in the soil. Despite the progress already achieved, combined efforts between plant breeding and biotechnology are needed to develop cultivars that will be more efficient in the use of phosphorus.

References

Ahsan M, Wright D, Virk DS (1996) Genetic analysis of salt tolerance in spring wheat (*Triticum aestivum* L.). Cereal Res Commun 24:353–360

Anghinoni, I. (2006). Phosphorus forms and availability in Brazilian tropical and subtropical soils under management systems. In: Proceedings of 3rd International Symposium on Phosphorus Dynamics in soil-Plant Continuum, Uberlândia, Sete Lagoas: Embrapa Milho e Sorgo, pp 13–14

Baligar VC, Fageria NK, He ZL (2001) Nutrient use efficiency in plants. Commun Soil Sci Plant Anal 32(1–8):921–950

Barriga BP, Proschle AA (1996) Herencia del contenido y de la eficiencia de la utilizacion del fosforo en trigo. Agro sur 12(1):43–49

Batten GD, Khan MA, Cullis BR (1984) Yield responses by modern wheat genotypes to phosphate fertilizer and their implications for breeding. Euphytica 33:81–89

Bolanos J, Edmeades GO (1993) Eight cycles of selection for drought tolerance in lowland tropical maize. II. Responses in reproductive behavior. Field Crops Res 31:253–268

Borlaug N (2000). The green revolution revisited and the road ahead. The Norwegian Nobel Institute, Oslo (special 30th anniversary lecture)

Chaubey CN, Senadhira D, Gregorio GB (1994) Genetic analysis of tolerance for phosphorous deficiency in rice (*Oryza sativa* L.). Theor Appl Genet 89(2–3):313–317

Chen J, Xu L, Cai Y, Xu J (2009) Identification of QTLs for phosphorus utilization efficiency in maize (*Zea mays* L.) across P levels. Euphytica 167(2):245–252

Cicarelli DM, Furlani AMC, Dechen AR, Lima M (1998) Genetic variation among maize genotypes for phosphorus-uptake and phosphorus-use efficiency in nutrient solution. J Plant Nutr 21:2219–2229

Dalal RC (1977) Soil organic phosphorus. Adv Agron 29:83–117

Dechen, AR, Nachtigall, RG (2007) Elementos requeridos à nutrição de plantas. In: Novais RF,Alvarez VVH, Barros NF, De Fontes , Cantarutti, RLF RB, Neves JCL (eds) Fertilidade do solo. Viçosa, MG: Sociedade Brasileira de Ciência do Solo, pp 91–132

Fageria NK, Baligar VC (1993) Screening crop genotypes for mineral stresses. In: Proceedings of theworkshop on adaption of plants to soil stresses, University of Nebraska, Lincoln, pp 142–162 (INTSORMIL Publications, 94-2)

FAO (2010) How to feed the world in 2050. Roma, Disponível em. http://www.fao.org/fileadmin/templates/wsfs/docs/expert_paper/How_to_Feed_the_World_in_2050.pdf. Acesso em: 17 Mar 2011
Furlani AMC, Lima M, Nass LL (1998) Combining ability effects for P- efficiency characters in maize grown in low P nutrient solution. Maydica 43:169–174
Furtini IV (2008) Implicações da seleção no feijoeiro efetuada em ambientes contrastantes em níveis de nitrogênio. Dissertação de Mestrado em Genética e Melhoramento de Plantas, Universidade Federal de Lavras, Lavras, p 67
Gorz HJ, Haskins FA, Pedersen JF, Ross WM (1987) Combining ability effects for mineral elements in forage sorghum hybrids. Crop Sci 27:216–219
Hisinger P, Jailard B, Le Cadre-Bartgélémy E, Plassard C., Drevon JJ (2006) The roots of phosphorus acquisition efficiency in crops In: Proceedings of the international symposium on phosphorus dynamics in the soil–plant continuum, 3., 2006, Sete Lagoas, Embrapa Milho e Sorgo, pp 75–76
Hyland C, Ketterings Q, Dewing D, Stockin K, Czymmek K, Albrecht G, Geohring L (2005) Phosphorus basics: the phosphorus cycle, Cornell University, Cornell (Agronomy Fact Sheet, 12). Disponível em. http://nmsp.cals.cornell.edu/publications/factsheets/factsheet12.pdf. Acesso em: 16 Mar 2011
Loughman BC, Roberts SC, Goodwin-Bailey CI (1983) Varietal differences in physiological and biochemical responses to changes in the ionic environment. Plant Soil 72:245–259
Moll RH, Kamprath EJ, Jackson WA (1982) Analysis and interpretation of factors which contribute to efficiency of nitrogen utilization. Agron J 74:562–564
Murrel TS, Fixen PE (2006) Improving fertilizer P effectiveness: challenges for the future. In: Proceedings of the international symposium on phosphorus dynamics in the soil-plant continuum, 3., 2006, Embrapa Milho e Sorgo, Sete Lagoas, pp 150–151
Ni JJ, Wu P, Senadhira D, Huang N (1998) Mapping QTLs for phosphorus deficiency tolerance in rice (*Oriva sativa* L.). Theor Appl Genet 97:1361–1369
Novais RF, Smyth TJ (1999) Fósforo em solo e planta em condições tropicais. UFV, Viçosa, p 399
Ochoa IE, Blair MW, Lynch JP (2006) QTL analysis for adventitious root formation in common bean under contrasting phosphorus availability. Crop Sci 46:1609–1621
Parentoni SN (2008) Estimativas de efeitos gênicos de diversos caracteres relacionados à eficiência e resposta ao fósforo em milho tropical. Tese (Doutorado)—Escola Superior de Agricultura Luiz de Queiroz, Piracicaba, p 207
Parentoni SN, Alves VMC, Milach SK, Cançado GMA, Bahia Filho AFC (2001) Melhoramento para tolerância ao alumínio como fator de adaptação a solos ácidos. In: Nass LL, Valois ACC, Melo IS, de Valadares-Inglis MC (eds) Recursos genéticos e melhoramento-plantas. Fundação MT, Rondonópolis, pp 783–851
Parentoni SN, De Souza Júnior CL, Alves VMC, Gama EEGE, Coelho AM, de Oliveira AC, Guimaraes CT, de Vasconcelos MJV, Pacheco CAP, Meirelles WF, De Magalhaes JV, Guimaraes LJM, da Silva AR, Mendes FF, Schaffert RE (2010) Inheritance and breeding strategies for phosphorus efficiency in tropical maize (*Zea mays* L.). Maydica 55(1):1–15
Parentoni SN, Souza Júnior CL (2008) Phosphorus acquisition and internal utilization efficiency in tropical maize genotypes. Pesquisa Agropecuária Brasileira 43(7):893–901
Parentoni SN, Vasconcelos CA, Alves VMC, Pacheco CAP, Santos MX, Gama EEG, Meirelles WF, Correa LA, Pitta GVE, Bahia Filho AFC (2000) Eficiência na utilização de fósforo em genótipos de milho. In: Congresso Nacional de Milho e Sorgo, 23., 2000, Uberlândia. A inovação tecnológica e a competitividade no contexto dos mercados globalizados: resumos. Sete Lagoas: ABMS: Embrapa Milho e Sorgo; Uberlândia: Universidade Federal de Uberlândia, p 92
de Resende MDV (2007) Matemática e estatística na análise de experimentos e no melhoramento genético. Embrapa Florestas, Colombo, p 362
Su J, Li H, Li B, Jing R, Tong Y, Li Z (2009) Detection of QTLs for phosphorus use efficiency in relation to agronomic performance of wheat grown under phosphorus sufficient and limited conditions. Plant Sci 176:824–836

Viana JHM, Resende ÁV, Guimarães CT, Parentoni SN (2009) Avaliação da variabilidade espacial do solo em experimentos de eficiência nutricional em milho, conduzidos em área com baixos teores de nutrientes: um estudo de caso. Sete Lagoas: Embrapa Milho e Sorgo, p 19. (Embrapa Milho e Sorgo. Boletim de Pesquisa e Desenvolvimento, 18)

Viana JH, Gomide RL, Albuquerque PEP, Durães, FOM, Andrade CDLT (2007) Protocolos para estabelecimento e caracterização de sítio específico experimental. Sete Lagoas: Embrapa Milho e Sorgo. p 6 (Embrapa Milho e Sorgo. Circular Técnica, 95)

Wang X, Shen J, Liao H (2010) Acquisition or utilization, which is more critical for enhancing phosphorus efficiency in modern crops. Plant Sci 179:302–306

Wissuwa M, Wegner J, Ae N, Yano M (2002) Substitution mapping of *Pup1*: a major QTL increasing phosphorus uptake of rice from a phosphorus-deficient soil. Theor Appl Genet 105:890–897

Wissuwa M, Yano M, Ae N (1998) Mapping of QTLs for phosphorus-deficiency tolerance in rice (*Oriva sativa* L.). Theor Appl Genet 97:777–783

Zhang D, Cheng H, Geng L, Kan G, Cui S, Meng Q, Gai J, Yu D (2009) Detection of quantitative trait loci for phosphorus deficiency tolerance at soybean seedling stage. Euphytica 167(3):313–322

Chapter 6
Breeding for Water Use Efficiency

Marcelo de Almeida Silva, Claudiana Moura dos Santos, Carlos Alberto Labate, Simone Guidetti-Gonzalez, Janaina de Santana Borges, Leonardo Cesar Ferreira, Rodrigo Oliveira DeLima and Roberto Fritsche-Neto

Abstract Among environmental factors, drought is the principal factor that limits global agricultural production. Many global agricultural regions are already suffering serious problems with water shortage and this scenario suggests that greater efforts are required in the development of alternatives for sustainable agriculture, such as the selection of cultivars that are efficient in the use of water. Water use efficiency (WUE) is the most important component of drought adaptation. In this chapter, WUE will be emphasized because of its greater importance and the possibility of obtaining genotypes that respond to environmental improvement or to the availability of water when it occurs. For this purpose

M. de Almeida Silva (✉) · C. M. dos Santos
Universidade Estadual Paulista, Sao Paulo, Brazil
e-mail: marcelosilva@fca.unesp.br

C. M. dos Santos
e-mail: claudianabio@hotmail.com

C. A. Labate · S. Guidetti-Gonzalez · J. de Santana Borges · L. C. Ferreira
Universidade de São Paulo, Sao Paulo, Brazil
e-mail: calabate@esalq.usp.br

S. Guidetti-Gonzalez
e-mail: sguidett@gmail.com

J. de Santana Borges
e-mail: jsborges@gmail.com

L. C. Ferreira
e-mail: ferreira.leonardocesar@gmail.com

R. O. DeLima · R. Fritsche-Neto
Universidade Federal de Viçosa, Minas Gerais, Brazil
e-mail: rodrigoodelima@gmail.com

R. Fritsche-Neto
e-mail: roberto.neto@ufv.br

R. Fritsche-Neto and A. Borém (eds.), *Plant Breeding for Abiotic Stress Tolerance*,
DOI: 10.1007/978-3-642-30553-5_6,

the following topics will be addressed: germplasm, genetic variability, relationship between characteristics, inheritance and maternal effect, oxidative stress, stress induction, selection strategies and breeding methods, and biotechnology applied to the breeding of WUE.

Keywords Water deficiency · Improved crops · Productivity · Abiotic stress · Selection strategies · Biotechnology

6.1 Introduction

Food shortages that plague the world are worrisome because agricultural production has not increased along with the increase in population and purchasing power of many countries. Among environmental factors, drought is the principal factor that limits global agricultural production (Manavalan et al. 2009). Events that cause water deficiency in plants have increased over the last decades and are likely associated with climate changes linked to global warming and the *El Niño* phenomenon (Hoerling and Kumar 2003).

Predictions estimate that global warming will be responsible for a 20 % increase in water scarcity, not only in drought-prone areas but also in various tropical and subtropical ecosystems. According to the scenario outlined by specialists, the cultivation areas of the large crop-producing regions will be practically reduced by half when the average temperature of the earth reaches 5.8 °C above the current temperature; this situation is predicted to occur within 50–100 years (Tilman et al. 2001). Predictions estimate that by the year 2020, approximately, one-third of the global population will be living in regions with water shortages, specifically aggravating the use of water in agriculture, which will become increasingly restricted. Moreover, the occurrence of poorly distributed and torrential rains, droughts, and heat waves will become more frequent and more severe and affect crop development and yield.

Many global agricultural regions are already suffering serious problems with water shortage, including China, India, and the USA; these countries represent three large global grain producers (Zhao et al. 2010). Grain yield losses of approximately 50 % have been observed for species such as corn, rice, wheat, and soy (Cha-Um et al. 2010). In Brazil, for example, producing regions in the South (which are responsible for more than 40 % of the national soy production) lost over 25 % of their production in recent harvests due to drought (Nepomuceno et al. 2009).

In forest systems, water use efficiency has an important relationship with wood production and water management. According to studies of global changes, water use efficiency is related to the carbon and water cycles in terrestrial vegetation. Thus, a future increase in global CO_2 concentrations is anticipated (Winter et al. 2001).

Another important question relates to the increasing pressure on farming due to its environmental impact. Among environmental factors, water deserves special attention. An estimated 70 % of the world's drinking water is used for irrigation (Fao 2007). Thus, one could easily predict that the use of water for irrigation will become increasingly costly.

This scenario suggests that greater efforts are required in the development of alternatives for sustainable agriculture, such as the selection of cultivars that are efficient in the use of water (Amudha and Balasubramani 2011).

6.2 Water Use Efficiency

Water use efficiency (WUE) is the most important component of drought adaptation. This parameter and its relationship to yield are often confused with drought tolerance. However, as discussed in Chapter 2, the strategies and the obtained materials are different in the selection of tolerance and WUE. In this chapter, WUE will be emphasized because of its greater importance and the possibility of obtaining genotypes that respond to environmental improvement or to the availability of water when it occurs.

Plants can be evaluated with regards to WUE from both the physiologic and agronomic point of view. Physiologically, WUE is the relationship between the CO_2 photosynthetic assimilation rate (A) and the plant's transpiration rate (Bänziger et al. 2006):

$$\mathrm{WUE} = \frac{(\mathrm{P_A} - \mathrm{P_I})}{1.6(\mathrm{VP_I} - \mathrm{VP_A})}$$

where

WUE: is the water use efficiency, physiologically,
P_A: partial pressure of CO_2 in the air,
P_I: partial pressure of CO_2 inside the leaf,
VP_I: vapor pressure of the water inside the leaf and
VP_A: vapor pressure of the water in the air.

Agronomically, WUE_a is the relationship between the dry mass produced and the volume of water used in the cycle (precipitation plus irrigation water) in the cultivated area (Jones 1993):

$$\mathrm{WUE} = \frac{\mathrm{GY}}{V}$$

where

WUE: water use efficiency in agronomic terms,
GY: grain yield or dry mass yield and
V: total water volume used in the cycle by the culture.

6.3 Germplasm, Genetic Variability and Relationship Between Characteristics

Under drought conditions, alterations occur in cell cycle and division, in membranes, in cell wall architecture, in metabolism, in the accumulation of osmotically active substances, osmolytes, and osmo-protectors and in carbohydrate metabolism (Bartels and Sunkar 2005). Such changes require the adjustment of metabolic pathways to achieve a new homeostatic state (Mittler 2006). These changes induce biochemical and physiological responses that include stomatal closure, a reduction in cellular growth and photosynthesis and the activation of respiration (Yamaguchi-Shinozaki and Shinozaki 2007).

Species differ greatly in the ability to overcome water deficiency. In sensitive species, a reduction in tissue hydration affects physiological processes. In contrast, tolerant species have physiological and metabolic properties that enable the plant to maintain a high degree of tissue hydration, even with limited water supply (Bartels and Sunkar 2005).

In maize, for example, substantial genetic variation exists in the grains yield and in secondary characteristics under water deficiency conditions. Bolaños and Edmeades (1993) have evaluated 208 maize testcrosses in Zimbabwe under both optimum and deficient soil water availability conditions. A drastic reduction in grain and a strong genotype x water availability interaction was observed, indicating that specific environments affect selection (Table 6.1) because the selected material in ideal cultivation conditions differed from those selected in water deficiency conditions.

In the same study, average yields of approximately 108 kg ha^{-1} $year^{-1}$ in grain yield could be obtained after eight cycles of recurrent selection under water deficiency conditions (Bolaños and Edmeades 1993). This observation again demonstrates that genetic variation exists in maize under water deficiency conditions. According to the same authors, these yields are due to greater synchrony in flowering, smaller tassel size, increased harvest index, increased *stay green,* and reduced root length density but not to changes in water absorption. Thus, when thinking about genetic improvements for water deficiency conditions, secondary conditions should always be considered in the adopted strategy.

Although water deficiency causes variability and reduced production, an opposite effect occurs in secondary characteristics, such as in the flowering range (Bänziger et al. 2006). This opposite effect is demonstrated in the results obtained by Ribaut et al. (1996). The authors evaluated 234 maize genotypes under various degrees of water deficiency and observed that the estimate of heritability in the flowering range was greater in deficiency conditions than in the optimal environment.

Few completed studies have directly estimated the genetic variability in WUE. Generally, WUE is measured through its components or related characteristics, such as specific leaf area, isotopic carbon discrimination, leaf chlorophyll level (SPAD) and others. Lal et al. (2006) have evaluated 36 forage peanut genotypes

Table 6.1 Grain yield (t ha^{-1}) of 208 maize testcrosses evaluated under two water conditions in Zimbabwe

Testcrosses	Culture conditions	
	Optimum	Water deficiency
Average of the 208	10.06 [a]	2.84 [a]
Least grain yield	5.85	0.51
Greatest grain yield	14.00	5.27

[a] Significant at 5 % level by the F test

Table 6.2 Summary of the variance analysis of characteristics related to WUE, including leaf chlorophyll level (SPAD), specific leaf area (SLA), harvest index (HI) and isotopic discrimination of Carbon 13 (^{13}C), in 36 forage peanut genotypes

Source of variation	GL	Average square			
		SPAD	SLA (cm g$-^{1}$)	HI	^{13}C
Repetition	1	79.98	0.36	143.31	2.77
Genotypes	35	43.06 [a]	516.86 [a]	195.42 [a]	0.82 [a]
Residue	35	2.84	104.13	33.19	0.30

[a] Significant at the 1 % in the F test, respectively
Source Adapted from Lal et al. (2006)

with regards to characteristics related to WUE (Table 6.2). According to the authors, genotypic variability exists in these four components, but the selection of more efficient genotypes for water use is still possible from these components.

As mentioned earlier, a series of possible morphophysiologic adaptations permits the plant to overcome and produce at a satisfactory rate under conditions of water stress (Xoconostle-Cazares et al. 2010). Some examples of these adaptations include decreased plant growth, reduced leaf area (Bastos et al. 2011) and increased growth of the root system (Varshney et al. 2011). Other strategies have been demonstrated, such as the rolling of leaves, floral abscission, and alterations in cuticle permeability.

In water deficiency environments, secondary characteristics can often assist in the selection of more productive or efficient genotypes in water use. This utilization of secondary characteristics is due to the fact that secondary characteristics become increasingly necessary in the selection of more efficient water use genotypes when compared with grain yield or WUE. Through secondary characteristics, the influence of water deficiency on the culture can be quantified and, if observed before maturity, can still be used as early selection criteria (Bänzinger et al. 2000).

It is important to emphasize that water deficiency influences the correlation estimates between secondary characteristics. This influence has been demonstrated in the case of maize by Ribaut et al. (1997). The authors observed that the magnitudes of correlations were greater under water deficiency conditions than under the optimum culture condition (Table 6.3). Moreover, in water deficiency, the estimate of correlation between the flowering interval and grain yield was

Table 6.3 Estimates of phenotypic correlation between grain yield (GY) and secondary characteristics, such as flowering interval (ASI), weight of 1,000 grains (W1000), number of spikes (NS), and number of grains per spike (NGS), in maize evaluated in optimum and water deficiency environments

Characteristics[1/]	ASI	GY	NS	NGS	W1000
ASI	1	−0.07 ns	−0.15[a]	−0.17 [b]	−0.17 [b]
GP	−0.39 [b]	1	−0.50 [b]	0.88 [b]	−0.22 [b]
NS	−0.55 [b]	0.64 [b]	1	0.56 [b]	−0.17 [b]
NGS	−0.43 [b]	0.94 [b]	0.70 [b]	1	−0.26 [b]
W1000	−0.06 ns	0.50 [b]	0.09 ns	0.20 [b]	1

[a, b, ns] Significant at the 1 % and 5 % levels and not significant at the 5% level in the t-test, respectively.[1/] The estimates below the diagonal line refer to the water deficiency environment and above the diagonal line refer to the optimum environment
Source Adapted from Ribaut et al. (1997)

Table 6.4 Contribution of dry root weight (DRW), specific leaf area (SLA), and chlorophyll level (SPAD) for WUE in forage peanut at three levels of water availability: without water deficiency (WD), with moderate deficiency (MD) and severe deficiency (SD)

Sources of variation	Explained by regression (%)		
	WD	MD	SD
Regression	59.39	96.50 [b]	72.31[a]
DRW	54.7 [b]	85.37 [b]	18.35[a]
SLA	4.48[a]	10.24[a]	52.65[a]
SPAD	0.21	0.89	1.30[a]

[a, b] Significant estimates at the 1 % and 5 % significance levels by the t-test, respectively
Source Adapted from Songsri et al. (2009)

significant and negative. This correlation is due to the abortion of ovules and, consequently, lower number of grains on the spikes and lower yield. Thus, the flowering interval is a secondary characteristic that must be considered in the breeding of maize for soil water deficiency conditions.

Other characteristics that can assist in the selection of more efficient water use genotypes involve the root system. Songsri et al. (2009) have studied the contribution of dry root weight (DRW), specific leaf area (SLA) and relative chlorophyll level (SPAD) in WUE in 11 forage peanut genotypes in the following three levels of water availability: without deficiency (WD), moderate deficiency (MD) and severe deficiency (SD). The authors observed an increased contribution of SLA and SPAD in WUE with an increased water deficiency level, suggesting that these characteristics can be good indicators in the selection of more efficient water use genotypes (Table 6.4). With regards to DRW, no linear association between DRW and water deficiency levels was observed.

Multiple studies of physiological characteristics and their interactions with grain yield in plants with water deficiency have been proposed. The selection criteria in these studies include minor decreases in the stomatal conductance and photosynthetic rate in soy (Gilbert et al. 2011), sugar cane (Machado et al. 2009), maize (Wang et al. 2008) and wheat (Monneveux et al. 2006). The opening of stomata may be regulated such that their closing is partial at a determined level of water deficiency, leading to increased efficiency in water use (Liu et al. 2005).

Another physiological characteristic, the maximum quantum efficiency of photosystem II (Fv/Fm), in which the ratio reflects the maximum efficiency when the light absorbed by the photosystem II antenna complex is converted into chemical energy (Maxwell and Johnson 2000), has been shown to be a good reference for diagnosing the photosynthetic system integrity in various species, such as maize (Zou et al. 2009), rice (Cha-Um et al. 2010), sugar cane (Silva et al. 2007) and wheat (Paknejad et al. 2007), under stress conditions and as a triage method for the selection of genotypes that are efficient in the use of water.

Regulation mediated by abscisic acid and by stomatal closing could also represent good parameters. The stomata are cells that are highly specialized in gas exchange and lose water by high leaf transpiration. For the plant's adaptation to drought, the stomata close during periods of stress. This stomata closing implies accumulation of gas, such as CO_2, which diminishes photosynthesis (Lawlor and Tezara 2009). The result is energy disequilibrium, in which the electron flow occurs in the direction of oxygen, creating reactive oxygen species and oxidative processes. Consequently, photochemical damage can occur in photosystem II (Noctor and Foyer 2005).

The accumulation of metabolites can also be important for the plants to adapt to environmental stresses. Among the metabolic alterations related to water deficiency, the accumulation of solutes is emphasized, such as glycine-betaine, proline, soluble amino acids, proteins, and sugars (Farhad et al. 2011). These solutes can accumulate under low water potential conditions, protecting the cells against dehydration by osmotic adjustment (Zhou and Yu 2010).

Among these solutes, proline is the most frequently studied. This metabolite acts as a mediator of osmotic adjustments, protects the integrity of the plasma membrane, provides a source of carbon and nitrogen and removes reactive oxygen species (Valliyodan and Nguyen 2006). In wheat (Vendruscolo et al. 2007) and other cultivated species, a positive relationship has been observed between the relative water content, the photosynthetic rate, the proline level and yield under water deficiency conditions.

The accumulation of soluble sugars and other macromolecules, such as LEA (*Late Embryogenesis Abundant*) proteins, in plants with water deficiency also correlates with WUE. Trehalose is highlighted among the sugars as a good indicator of WUE in studies of wheat (El-Bashiti et al. 2005) and rice (Soren et al. 2010).

6.4 Inheritance and Maternal Effect

WUE is a complex and determinative characteristic of the grains yield, principally in conditions of water deficiency (Blum 2005). The limited improvement in WUE in different species could be due to the fact that this characteristic is controlled by multiple genes and the strong interaction between their expression and the water availability levels (Ashraf 2010). Therefore, the knowledge of genetic mechanisms

Table 6.5 General (GCA) and specific (SCA) analysis of variance of combining ability and reciprocal effect of the WUE, chlorophyll level (SPAD), specific leaf area (SLA), harvest index (HI), and isotopic discrimination of carbon 13 (^{13}C) in forage peanuts

Sources of variation	GL	SPAD	SLA (cm g^{-1})	HI	^{13}C
GCA	5	125.40 [b]	1192.04 [b]	502.99 [b]	0.95 [b]
SCA	15	7.22 [b]	73.33 ns	29.71 ns	0.34[a]
Reciprocal	15	1.19 ns	132.32[a]	30.61 ns	0.30[a]
Residue	35	52.02	16.59	16.59	0.15

[a, b, ns] Significant at the 1 % and 5 % levels and lack of significance at the 5 % level by the F-test, respectively

Source Adapted from Lal et al. (2006)

involved in the water deficiency response is important because it allows for the identification of the genes expressed under these conditions and the manipulation of this information to obtain cultivars that are more efficient in water use.

With this objective, Derera et al. (2008) have studied the genetic control of grain yield and water deficiency. Regarding grain yield, additive genetic effects predominate under water deficiency conditions, but under normal cultivation conditions, additive and non-additive genetic effects are important in the control of this characteristic. Therefore, in water deficiency characteristics, WUE can be selected based on the performance of lineages that comprise the hybrids.

In the same study, the authors observed a significant maternal effect in the grains yield under conditions of water deficiency. Thus, significant differences exist in the choice of the female progenitor in determined crosses for grain production.

In another study, Lal et al. (2006) identified the genetic control of WUE components in forage peanuts. According to these authors, the SPAD and isotopic carbon discrimination is affected by the influence of additive and non-additive genetic effects. With regards to the SLA and the HI, only the additive genetic effects were important in the control of its expression (Table 6.5). Moreover, the maternal effect in the SLA and the isotopic carbon discrimination suggests an important role for the female pro in the improvement of these two characteristics. Therefore, distinct breeding strategies exist to increase WUE (through its components), depending on the component; the selection can be made in lines *per se* or in hybrid combinations.

Significant maternal effects in the physiological parameters related to WUE are expected; therefore, photosynthetic structure and energy production display extra-chromosomal inheritance. Nevertheless, studies that quantify the effect of choosing the progenitor in the breeding process are rare for this type of abiotic stress.

6.5 Oxidative Stress

One of the principle effects of plant dehydration is the production of reactive oxygen species (ROS), such as singlet oxygen, superoxide radicals, hydroxyl radicals, and hydrogen peroxide (Noctor and Foyer 2005). ROS are normally produced in various metabolic pathways and in different plant organelles (such as within the chloroplasts, in the mitochondria and in peroxisomes, particularly during photosynthesis and respiration) in very low quantities (Apel and Hirt 2004). Reports have indicated that the exposure of plants to various types of environmental stress, such as drought, can intensify the production of ROS, which results in increased harmful oxidation of the cellular components, leading to oxidative stress in the plants (Noctor and Foyer 2005).

The antioxidant system enzymes are extremely sensitive to environmental stress conditions and serve as signalers of the increased or decreased expression of genes in plants submitted to water deficit conditions. These enzymes can be used as molecular markers of drought tolerance (Badawi et al. 2004).

Plant survival mechanisms under stress conditions are dependent on different responses, including the capacity of the plant to maintain high antioxidant levels, whether enzymatic or non-enzymatic (Apel and Hirt 2004). Plants have multiple genes that are encoded based on a key enzyme, resulting in different isoforms of antioxidant enzymes, such as superoxide dismutase (SOD, EC 1.15.1.1), catalase (CAT, EC 1.11.1.6), ascorbate peroxidase (APX, EC 1.11.1.11), glutathione peroxidase (GPX, EC 1.11.1.9), peroxidase (POX, EC 1.11.1.7) and glutathione reductase (GR., EC 1.8.1.7). These enzymes are extremely sensitive to environmental stress conditions and serve as signals of the stress. Therefore, the role of antioxidant enzymes in stress situations is to control the accumulation of ROS and, thus, to limit oxidative damage (Asada 2006).

Various studies have shown alterations in the activity of antioxidant enzymes in plants in response to water deficiency, suggesting that the increase in these activities can be directly related to stress tolerance. For example, the activity of SOD and its FeSOD isoform increases the WUE in potato genotypes under conditions of water deficiency (Huseynova et al. 2010). In turn, canola exhibits increases in CAT and APX (Omidi 2010), while wheat shows increases in CAT1 and CAT2 isoforms (Luna et al. 2005).

6.6 Stress Induction: Duration, Intensity, and Uniformity

The management of stress is crucial for the success of breeding programs for abiotic stress conditions, principally water deficiencies, which is a factor that can interact with the others. In this case, if the stress induction were accentuated, the plant might not be able to complete its life cycle or, furthermore, not produce grains or biomass in reasonable quantities to distinguish the genotypes.

Water potential is considered to be a trustworthy parameter for measuring the plant's response to water deficiency. This parameter varies with the type of plant and the environmental conditions. According to Hsiao (1973), the majority of plant responses to water deficiency (such as cell growth, cell wall and protein synthesis and enzymatic activity) are influenced by water potential reductions of less than -1.5 MPa.

Improving WUE has been the focus of many research centers. The majority of studies are conducted under field conditions, although some investigators are utilizing greenhouses for this purpose. However, these studies have different objectives. In the field, the purpose is to select the genotypes that are most efficient in water use, but in greenhouses, the purpose is to understand the mechanisms that contribute to WUE under water deficiency conditions.

Duration, intensity and uniformity are factors that must be considered to establish adequate stress management (Bänziger et al. 2000). The duration must coincide with the critical stages of plant development and involve characteristics that can be modified by genetic improvement. The intensity must be severe enough to affect the important yield characteristics, permitting differentiation in germplasm when plants are submitted to optimal conditions of water availability and to water deficiency. Finally, the handling must be uniform, in time and space, such that genetic variance can be easily observed, allowing for genetic gains.

The key to breeding under water deficiency conditions in the field is to manage the stress well. In the case of water deficiency, improvements are made by conducting experiments partially or entirely during the drought season and with the stress managed by irrigation. The objective of these experiments is not to randomly simulate water deficiency conditions but to clearly simulate normal and water deficiency culture conditions.

To study WUE in field experiments in maize, for example, the water deficiency is applied in two stages: in the pre-flowering and grain-filling stages. The stress applied in the pre-flowering stage has the objective of slowing male flowering and causing abortion of grains and spikes. The water supply is interrupted 15 days before flowering and returned at the beginning of grain filling. Water deficiency applied to grain filling affects grain weight due to a reduction in photosynthesis during this phase. In this case, the grain yield must be reduced by approximately 50 % relative to the optimum water conditions (Bolaños and Edmeades, 1993).

In early stage experimental evaluations of the plant, large polyethylene boxes filled with substrate formed from a mixture of sand and soil are normally utilized. These boxes are subdivided into smaller units, and each unit receives a plant. The plants are evaluated at the V8 stage, when, in the optimum environment, the plants generally receive quantities of water that are sufficient to restitute the amount lost by evaporation. To simulate water deficiency, the plants receive half of the quantity of water of that which the plants receive under the optimum conditions (Hund et al. 2009).

In another type of experiment in which a small number of genotypes are evaluated, the plants are cultivated in 20 dm^3 polyethylene pots. These pots are filled with soil, with two plants per pot, from seeding to the final cycle. The plants are evaluated with regards to physiological characteristics, such as photosynthetic activity, chlorophyll level, intrinsic water use efficiency and others. Water replacement is performed based on readings obtained with a sensor, and the water is replaced to field capacity. In the pre-flowering stage, water deficiency is imposed such that only 50 % of the total water applied in the optimum environment is applied (Magalhães et al. 2009). In both situations, fertilization in the substrate is performed based on the technical recommendation for the studied culture after chemical analysis of the substrate.

6.7 Selection Strategies and Breeding Methods

Similarly to other characteristics, those that present high heritability and genetic control (principally of an additive type) in WUE can be selected by individual performance (by the performance *per se* of the lineages or populations). When the characteristic is of low heritability or presents heritability due to non-additive effect genes, genotypes should be selected based on the hybrid performance. Importantly, as previously mentioned, a significant maternal effect will cause differences in the choice of which will be the female progenitor in a determined cross.

As shown in this chapter, divergent results can be obtained in the breeding for WUE based on the level of water availability, the germplasm, and characteristics. Thus, it is important that each breeding program studies the WUE heritance in the germplasm under study. Other important studies to assist in selection refer to the identification and quantification of the genetic correlations between the water efficiency characteristics and between the agronomic characteristics. The goal is to understand whether it is possible to complete the indirect selection of low heritability characteristics and/or of easy evaluation, either due to experimental difficulties or due to the high cost of measurement.

Finally, quantifying the genotype x availability of water level interactions is important. Characteristics for selection should not be used based on the performance of the genotypes in only one environment because, as mentioned, genes are expressed differently based on the availability of water to the plant.

Similar to the breeding of other characteristics, WUE breeding methods are chosen based on the reproductive system of the species, the cultivar type that is desired, the heritability and the genetic control of the most important characteristics.

Population methods are based on recurrent selection and aim to gradually increase the frequency of favorable alleles for quantitative characteristics (through repeated selection cycles) without significantly reducing the genetic variability of the population. Recurrent selection can be divided into obtaining lineages, evaluating lineages in repeated experiments and recombination of the superior lineages to originate the following generation. The improved populations can be repeatedly used to begin a new recurrent selection cycle after the recombination of the selected and superior lineages (Bernardo 2002).

When WUE demonstrates genetic control principally by additive effects, intrapopulation recurrent selection methods can be used. However, when a heterosis effect exists (a non-additive effect), the interpopulation methods are faster and more appropriate for obtaining cultivars that are superiors in WUE.

6.8 Biotechnology Applied to the Breeding of WUE

6.8.1 *Quantitative Trait Loci*

Quantitative Trait Loci (QTL) mapping allows for improved understanding of the genetic control and inheritance of WUE and can indicate which selection strategy should be adopted. However, functional genomic approaches that provide genome-scale perspectives on the gene network related to water metabolism are important to identify key points that can be modified through biotechnology. QTLs of agricultural characteristics in water deficiency have been described in various cereal species.

Due to the importance of the flowering interval in WUE in maize, principally under water deficiency conditions, the CIMMYT (International Center for Maize and Wheat Improvement—Centro Internacional de Melhoramento de Milho and Trigo) has identified six QTLs for the flowering interval and grain yield components under these conditions. These QTLs explain 50 % of the phenotypic flowering interval variation and have been stable over multiple years and under different water deficiency levels (Ribaut et al. 1997).

In red-oak (*Quercus robur* L.), Brendel et al. (2008) have identified ten QTLs that can discriminate the carbon 13 isotope, one of which can explain greater than 20% of the phenotypic variation. Moreover, other QTLs have been identified for the chlorophyll level in the leaf colocalized with the QTL for WUE and the foliar area mass.

Recent advances in statistical genomics and molecular biology will improve the precision of identifying QTLs associated with WUE in the near future. However, to obtain consistent QTL results for WUE for the different water availabilities in the soil and from various characteristics, improvements in phenotyping will be important. For these improvements, large populations are used for mapping (evaluated at different locations and over multiple years) to observe the maximum number of characteristics related to WUE, and more accurate tools are used, such as network assays and mixed model methodology in the statistical analyses. Moreover, the characterization of QTLs identified by sequencing analysis will permit the identification of possible structural or regulatory genes controlling WUE during plant development and under different water availability conditions in the soil.

6.8.2 *Selection Assisted by Markers*

In the evaluation and selection based on the WUE phenotype and its components, methods that are destructive to the plants are generally used, which, depending on the genetic structure of the plant, does not allow for the plants that are identified as superior to be used in hybridizations. Consequently, reductions occur in the yields obtained by selection. Furthermore, some conventional breeding methods, such as recurrent selection, are very laborious and present low yields with the selection for time units, often becoming economically impracticable.

Some studies have already shown the potential of this technique. An example is the work by Eathington (2005), who obtained genetic gains in maize grain production with recurrent selection assisted by molecular markers two times greater than those obtained with recurrent phenotypic selection. In other studies, selection assisted by molecular markers for smaller intervals of flowering in maize was a powerful tool to increase WUE (Condon et al. 2004) and its components in water deficiency conditions (Ribaut and Ragot 2007).

6.8.3 Transgenics

Genetic transformation techniques in plants have been applied in the identification and transfer of genes responsible for drought tolerance or WUE. Genes that are expressed during stress are anticipated to be involved in cellular tolerance to dehydration based on their protection functions in the cytoplasm and in the cellular membrane, their alterations to promote water absorption in the cell, their control ions accumulation and their regulation of additional genes (Roy et al. 2011).

Many of these genes have been identified by microarray analysis of *Arabidopsis*; the products of these genes can be classified into two groups. The first group contains proteins that appear to directly act on tolerance to abiotic stress, such as chaperones, LEA proteins anti-freezing and mRNA ligation proteins, enzymes for the biosynthesis of osmolytes, aquaporins, ionic transporters (of sugars and amino acids), detoxifying enzymes, and proteases (Roy et al. 2011).

The second group consists of regulatory proteins and contains factors involved in the regulation or transduction of signals during the stress response. This group includes transcription factors, kinase and phosphatase proteins and enzymes that metabolize phospholipids and other signaling molecules, such as calmodulin ligating proteins (Yamaguchi-Shinozaki and Shinozaki, 2007).

A series of cultivars developed by transgenics for water deficiency tolerance have also been obtained through the ROS detoxification strategy in various species, including transgenics expressing enzymes with different isoforms involved in oxidative protection (Amudha and Balasubramani 2011). In transgenic tobacco, the differential expression of cytosolic cAPX in chloroplasts has been verified and has resulted in increased drought tolerance (Badawi et al. 2004).

In studies performed with sugar cane, Molinari et al. (2007) have observed greater tolerance to water deficiency in the genotypes that were transformed to increase the capacity to accumulate proline and increase P5CS enzyme activity (Δ1-pyrroline-5-carboxylate synthetase, which is an enzyme that belongs to the proline biosynthesis metabolic pathway). The authors also observed that proline acted more as an antioxidant agent than as an osmoprotector.

References

Amudha J, Balasubramani G (2011) Recent molecular advances to combat abiotic stress tolerance in crop plants. Biotechnol Mol Biol Rev 6:31–58

Apel K, Hirt H (2004) Reactive oxygen species: metabolism, oxidative stress, and signal transduction. Annu Rev Plant Biol 55:373–399

Asada K (2006) Production and scavenging of reactive oxygen species in chloroplasts and their functions. Plant Physiol 141:391–396

Ashraf M (2010) Inducing drought tolerance in plants: Recent advances. Biotechnol Adv 28:169–183

Badawi GH, Kawano N, Yamauchi Y, Shimada E, Sasaki R, Kubo A, Tanaka K (2004) Over-expression of ascorbate peroxidase in tobacco chloroplasts enhances the tolerance to salt stress and water deficit. Physiol Plant 121:231–238

Bänziger M, Edmeades GO, Beck D, Bellon M (2000) for drought and nitrogen stress tolerance in maize. CIMMYT Special Publication. CIMMYT, Mexico, p 68

Bänziger M, Setimela PS, Hodson D, Vivek B (2006) Breeding for improved drought tolerance in maize adapted to southern Africa. Agric Water Manag 80:212–224

Bastos EA, Nascimento SP, Silva EM, Freire Filho FR, Lúcio Gomide RL (2011) Identification of cowpea genotypes for drought tolerance. 2011. Revista Ciência Agronômica 42:100–107

Bartels D, Sunkar R (2005) Drought and salt tolerance in plants. Crit Rev Plant Sci 24:23–58

Bernardo R (2002) Breeding for quantitative traits in plants. Stemma Press, Woodbury, p 369

Bolaños J, Edmeades GO (1993) Eight cycles of selection for drought tolerance in lowland tropical maize. I. Responses in grain yield, biomass, and radiation utilization. Field Crops Res 31:233–252

Blum A (2005) Drought resistance, water-use efficiency, and yield potential—are they compatible, dissonant, or mutually exclusive? Aust J Agric Res 56:1159–1168

Brendel O, Thiec DL, Scotti-Saintagne C, Bodenes C, Kremer A, Guehl J (2008) Quantitative trait loci controlling water use efficiency and related traits in *Quercus robur* L. Tree Genet Genomes 4:263–278

Cha-Um S, Nhung NTH, Kirdmanee C (2010) Effect of mannitol- and salt-induced iso-osmotic stress on proline accumulation, photosynthetic abilities and growth characters of rice cultivars (*Oryza sativa* L. spp. *Indica*). Pak J Bot 42:927–941

Condon AG, Richards RA, Rebetzke GJ, Farquhar GD (2004) Breeding for high water-use efficiency. J Exp Bot 55:2447–2460

Derera J, Tongoona P, Vivek BS, Laing AD (2008) Gene action controlling grain yield and secondary traits in southern African maize hybrids under drought and non-drought environments. Euphytica 162:411–422

Eathington S (2005) Practical applications of molecular technology in the development of commercial maize hybrids. In: Proceedings of the 60th Annual Corn and Sorghum Seed Research Conferences. Washington, DC: American Seed Trade Association

El-Bashiti T, Seyin H, Ktem AO, Yü Cel M (2005). Biochemical analysis of trehalose and its metabolizing enzymes in wheat under abiotic stress conditions. Plant Science 169:47-54

FAO (2007) Current World Fertilizer Trends and Outlook to 2011/12. Food and Agriculture Organization of the United Nations, 2007

Farhad MS, Babak AM, Reza ZM, Hassan RSM, Afshin T (2011) Response of proline, soluble sugars, photosynthetic pigments and antioxidant enzymes in potato (*Solanum tuberosum* L.) to different irrigation regimes in greenhouse condition. Aust J Crop Siencie 5:55–60

Gilbert ME, Zwieniecki MA, Holbrook NM (2011) Independent variation in photosynthetic capacity and stomatal conductance leads to differences in intrinsic water use efficiency in 11 soybean genotypes before and during mild drought. J Exp Bot 62:1–13

Hsiao TC (1973) Plant responses to water stress. Ann Rev Plant Physiol 24:519–570

Hoerling M, Kumar A (2003) The perfect ocean for drought. Science 299:691–694

Hund A, Ruta, N, Liedgens M (2009). Rooting depth and water use efficiency of tropical maize inbred lines, differing in drought tolerance. Plant Soil 318–311–325

Huseynova IM, Suleymanov SY, Rustamova SM (2010) Response of photosynthetic apparatus and antioxidant defense systems in *Triticum aestivum* L. genotypes subjected to drought stress. Proceedings of ANAIS. (Biological Sciences) 65:49–59

Jones HJ (1993) Drought tolerance and water-use efficiency. In: Smith JAC, Griffiths H (eds) Water deficits, plant response from cell to community. BIOS Scientific Publishers, Oxford, pp 193–203

Lal CL, Hariprasanna K, Rathnakumar AL, Gor HK, Chikani BM (2006) Gene action for surrogate traits of water-use efficiency and harvest index in peanut (*Arachis hypogaea*). Ann Appl Biol 148:165–172

Lawlor DW, Tezara W (2009) Causes of decreased photosynthetic rate and metabolic capacity in water-deficient leaf cells: a critical evaluation of mechanisms and integration of processes. Ann Bot 103:561–579

Liu F, Anderson MN, Jacobson SE, Jensen CR (2005) Stomatal control and water use efficiency of soybean (*Glycine max* L. Merr.) during progressive soil drying. Environ Exp Bot 54:33–40

Luna CM, Pastori GM, Driscoll S, Groten K, Bernard S, Foyer CH (2005) Drought controls on H_2O_2 accumulation, catalase (CAT) activity and CAT gene expression in wheat. J Exp Bot 56:417–423

Machado RS, Ribeiro RV, Marchiori PER, Machado DFSP, Machado EC, Landell MGA (2009) Respostas biométricas e fisiológicas ao deficit hídrico em cana-de-açúcar em diferentes fases fenológicas. Pesquisa Agropecuária Brasileira 44:1575–1582

Magalhães PC, Souza TC, Albuquerque PEP, Karam D, Magalhães MM, Cantão FRO (2009) Caracterização ecofisiológica de linhagens de milho submetidas a baixa disponibilidade hídrica durante o florescimento. Revista Brasileira de Milho e Sorgo 8:223–232

Manavalan LP, Guttikonda SK, Tran LSP, Nguyen HT (2009) Physiological and molecular approaches to improve drought resistance in soybean. Plant Cell Physiol 50:1260–1276

Maxwell K, Johnson GN (2000) Chlorophyll fluorescence: a pratical guide. J Exp Bot 51:659–668

Mittler R (2006) Abiotic stress, the field environment and stress combination. Trends Plant Sci 11:15–19

Molinari HBC, Marur CJ, Daros E, Campos MKF, Carvalho JFRP, Bespalhok Filho JC, Pereira LFPP, Vieira LGE (2007) Evaluation of the stress-inducible production of proline in transgenic sugarcane (*Saccharum* spp.): osmotic adjustment, chlorophyll fluorescence and oxidative stress. Physiol Plant 130:218–229

Monneveux H, Rekika D, Acevedo E, Othmane Merah O (2006) Effect of drought on leaf gas exchange, carbon isotope discrimination, transpiration efficiency and productivity in field grown durum wheat genotypes. Plant Sci 170:867–872

Nepomuceno A L, Fuganti R, Rodrigues FA, Neumaier N, Farias JRB, Kanamori N, Marcelino C (2009) Estratégias moleculares para tolerância a seca em plantas. In: Lacerda CF, Gomes Filho E, Bezerra MA, Marques EC (orgs.). A fisiologia vegetal e os desafios para produção de alimentos e bioenergia. Fortaleza: UFC/EMBRAPA-CNPAT.

Noctor G, Foyer CH (2005) Redox homeostasis and antioxidant signaling: a metabolic interface between stress perception and physiological responses. Plant Cell 17:1866–1875

Omidi H (2010) Changes of proline content and activity of antioxidative enzymes in two canola genotype under drought stress. American Journal of Plant Physiology 5:338–349

Paknejad F, Nasri M, Moghadam HRT, Zahedi H, Alahmadi MJ (2007) Effects of drought stress on chlorophyll fluorescence parameters, chlorophyll content and grain yield of wheat cultivars. International Journal of Biological Science 7:841–847

Ribaut JM, Hoisington DA, Deutsch JA, Jiang C, González-de-Leon D (1996) Identification of quantitative trait loci under drought conditions in tropical maize. I. Flowering parameters and the anthesis-silking interval. Theoretical and Applied Genetics 92:905–914

Ribaut JM, Hoisington D, Edmeades GO, Huerta E, González-de-León D (1997) Changes in allelic frequencies in a tropical maize population under selection for drought tolerance.

In: Edmeades GO, Bänziger M, Mickelson HR, Peña-Valdivia CB. (eds.). Developing Drought and Low N-Tolerant Maize. Proceedings of a Symposium, March 25-29, 1996, CIMMYT, El Batán, México. D.F.: CIMMYT. pp 392–395

Ribaut JM, Ragot M (2007) Marker-assisted selection to improve drought adaptation in maize: the backcross approach, perspectives, limitations and alternatives. J Exp Bot 58:351–360

Roy B, Noren SK, Mandal AB, Basu AK (2011) Genetic engineering for abiotic stress tolerance in agricultural crops. Biotechnology 10:1–22

Silva MA, Jifon JL, Silva JAG, Sharma V (2007) Use of physiological parameters as fast tools to screen for drought tolerance in sugarcane. Braz J Plant Physiol 19:193–201

Songsri P, Jogloy S, Holbrook CC, Kesmala T, Vorasoot N, Akkasaeng C, Patanothai A (2009) Association of root, specific leaf area and SPAD chlorophyll meter reading to water use efficiency of peanut under different available soil water. Agric Water Manag 96:790–798

Soren KR, Kishwar A, Vandana T, Aruna T (2010) Recent advances in molecular breeding of drought tolerance in rice (*Oryza sativa* L.). Ind J Biotech 9:233–251

Tilman D, Fargione J, Wolff B, D'antonio C, Dobson A, Howarth R, Schindler D, Schlesinger WH, Simberloff D, Swackhamer D (2001) Forecasting agriculturally driven global environmental change. Science's Compass Rev 292:281–284

Varshney RK, Pazhamala L, Kashiwagi J, Gaur PM, Krishnamurthy L, Hoisington D (2011) Genomics and physiological approaches for root trait breeding to improve drought tolerance in chickpea (*Cicer arietinum* L.). Root Genomics 10:213–222

Valliyodan B, Nguyen HT (2006) Understanding regulatory networks and engineering for enhanced drought tolerance in plants. Curr Opin Plant Biol 9:1–7

Vendruscolo ACG, Schuster I, Pileggi M, Scarpim CA, Molinari HBC, Marur CJ, Vieira LGC (2007) Stress-induced synthesis of proline confers tolerance to water deficit in transgenic wheat. J Plant Physiol 164:1367–1376

Wang B, Li Z, Eneji AE, Tian X, Zhai Z, Li J, Duan L (2008) Effects of coronatine on growth, gas exchange traits, chlorophyll content, antioxidant enzymes and lipid peroxidation in maize (*Zea mays* L.) seedlings under simulated drought stress. Plant Prod Sci 11:283–290

Winter K, Aranda J, Garcia M, Virgo A, Paton SR (2001) Effect of elevated CO_2 and soil fertilization on whole-plant growth and water use in seedlings of a tropical pioneer tree, *Ficus insipida* Willd. Flora 196:458–464

Xoconostle-Cazares B, Ramirez-Ortega FA, Flores-Elenes L, Ruiz-Medrano R (2010) Drought tolerance in crop plants. Amer J Plant Physiol 5:241–256

Yamaguchi-Shinozaki K, Shinozaki K (2007) Gene networks involved in drought stress response and tolerance. J Exp Bot 58:221–227

Zhao J, Sun H, Dai H, Zhang G, Wu F (2010) Difference in response to drought stress among Tibet wild barley genotypes. Euphytica 172:395–403

Zhou Q, Yu B (2010) Changes in content of free, conjugated and bound polyamines and osmotic adjustment in adaptation of vetiver grass to water deficit. Plant Physiol Biochem 48:417–425

Zou J, Yu K, Zhang Z, Jiang W, Liu D (2009) Antioxidant response system and chlorophyll fluorescence in chromium (vi) treated *Zea mays* L. seedlings. Acta Biologica Cracoviensia—Series Botanica 51:23–33

Chapter 7
Breeding for Salinity Tolerance

Nand Kumar Fageria, Luís Fernando Stone and Alberto Baêta dos Santos

Abstract Soil salinity is a major factor adversely affecting crop yields worldwide. It is estimated that worldwide about 1 billion ha of land is affected by salinity. In addition, salinity problem is increasing at a rate of about 10% annually worldwide. Salinity can cause a combination of complex interactions that affect plant metabolism, susceptibility to injury or internal nutrient requirement. The negative interactions of salts with crop plants may reduce growth and consequently nutrient use efficiency. Management practices which can be adopted to reduce negative effects of salts on plant growth includes leaching salts from soil profile, use of amendments such as gypsum, and use of farmyard manners. However, use of salt-tolerant crop species or genotypes within species is a very attractive strategy to reduce cost of salt reclamation and environmental pollution. Although salt tolerance is relatively low in most crop species, it is encouraging that genetic variability exists not only among species but also among genotypes of same species. Salt-tolerant crop species are barley, cotton, oats, rye, triticale, sugar beet, guar, and canola or rapseed. Plant resistance responses include both salt avoidance (selective uptake or exclusion mechanisms and salt secretion, such as through salt glands) and salt tolerance (osmotic adjustment to maintain turgor pressure, tissue tolerance to specific toxic ions, e.g., Na and Cl, and tissue dehydration tolerance).

Keywords Absorption of nutrients · Genetic variability · Electrical conductivity · Salinity threshold · Dry matter or grain yield efficiency index · Breeding methods for salinity · Genetic markers · Quantitative trait loci (QLT) · Transgenic method

---

N. K. Fageria (✉) · L. F. Stone · A. B. dos Santos
National Rice and Bean Research Center of EMBRAPA, Caixa Postal 179, Santo Antônio de Goiás, GO 75375-000, Brazil
e-mail: fageria@cnpaf.embrapa.br

R. Fritsche-Neto and A. Borém (eds.), *Plant Breeding for Abiotic Stress Tolerance*,
DOI: 10.1007/978-3-642-30553-5_7, © Springer-Verlag Berlin Heidelberg 2012

7.1 Introduction

Soil salinity is a major factor which reduces crop production worldwide. Major areas affected by salts are located in North, central, and South Asia (Table 7.1). In South America, areas affected by salinity are about 85 million hectares, including saline and sodic soils. In Brazil, major soils affected by salinity are in the North-East region. However, soils affected by salts are also found in the state of Rio Grande do Sul and Pantanal of Mato Grosso (Ribeiro et al. 2009). These authors also reported that in Brazil soils affected by salts are about 160.000 km^2 or 2 % of the national territory. Major part of the soils affected by salinity are in the state of Bahia (44 % of total), followed by state of Ceará with 26 % of total area. Salinity in irrigated lands is determined by irrigation water quality, irrigation method, drainage, soil permeability, and water table. In Brazil, salinity is mostly prevailing in the dry area of nine states of the North-East. In some irrigated areas of North-East, salinity decreased yield of crops and farmers abandoned these lands. It is estimated that about 20 % of the irrigated areas of the North-East of Brazil are affected by salinity.

About 20 % of cultivated area and 33 % of irrigated area is affected by salinity worldwide and this area is mostly located in Asia (Rains and Goyal 2003, Ashraf and Foolad 2007). Worldwide about 1 billion ha of land is affected by salinity. According to Pessarakli and Szabolcs (1999), all continents have salinity problem, except Antarctica. In addition, salinity problem is increasing at a rate of 10 % annually (Szabolcs 1994). These data are alarming because world population from 7 billion people in 2011 to be expected to increase about more than 9 billion people in 2050 (Epstein and Bloom 2005). Major part of this population increase will be in the developing countries where food demand is higher. In this context, incorporation of land areas affected by salts in crop production have important role in future from social and economic point of view (Fageria et al. 2010). Furthermore, urbanization and industrialization will increase competition for fresh water (Rains and Goyal 2003). Hence, inadequate soil and water management will increase salinity problem worldwide (National Academy of Sciences 1999).

Soils affected by salts are defined as those which are adversely modified for growth and development of most crop species due to presence of soluble salts, exchangeable sodium or both in the rhizosphere (Soil Science Society of America 2008). Soluble salts normally present in salt affected soils are cations such as Ca^{2+}, Mg^{2+}, Na^+ and anions SO_4^{2-}, Cl^-, HCO_3^-, and sometimes K^+, CO_3^{2-}, and NO_3^-. Soils affected by salts are mostly common in arid and semi-arid regions due to low precipitation and high evaporation. These climatic conditions do not permit salt lixiviation from the soil profile and accumulate in quantity toxic for plant growth. Salinity problem also occurs in areas near sea shore due to flooding by sea water which contain high salts (Fageria et al. 2010). Use of inappropriate levels of fertilizers with inadequate management practices can create saline conditions even in humid conditions (Fageria et al. 2011a).

Table 7.1 Salt affected area in the world

Region	Area in 1.000 hectares		
	Saline	Sodic	Total
North America	6,191	9,564	15,755
Mexico and Central America	1,965	–	1,965
South America	69,410	59,573	128,983
Africa	53,492	26,946	80,438
South Asia	83,312	1,798	85,110
North and Central Asia	91,621	120,065	211,686
East Asia	19,983	–	19,983
Australia and New Zealand	17,359	339,971	357,330
Total	343,333	557,917	901,250

Source Adapted from Lal et al. (1989)

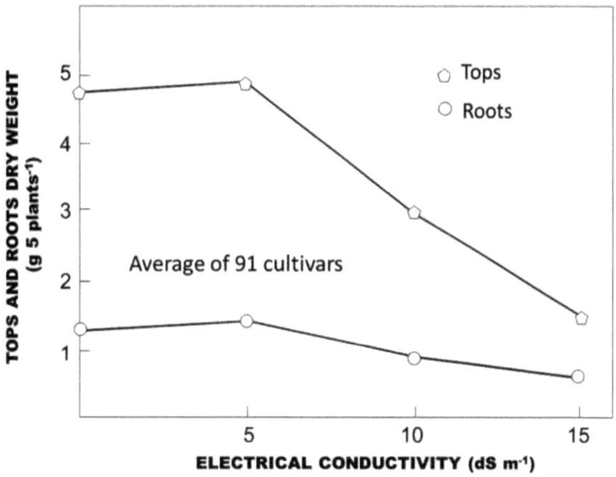

Fig. 7.1 Relationship between salinity levels and dry matter of shoots and roots of lowland rice. Adapted from Fageria (1992)

Negative effects of salinity on agriculture are a concern because it affects growth and development and yield of crop plants. Typically, decrease in growth of plants occurs linearly after attending threshold value of salinity. Leaves of salt affected plants are small and show dark green color in comparison with leaves of normal plants. Generally, salinity effects on plants do not show yellowing or discoloration such as nutritional deficiency. Salinity decreases water absorption but wilting symptoms of leaves rarely occur. Under saline conditions moderately low water potentials are always present and water potential changes are usually gradual. Plants are, therefore, hardened by the continual stress and are less apt to exhibit abrupt change in turgor. Plants also do not show symptoms such as marginal or tip burn of leaves, occur as a rule only in woody plants in which these symptoms indicate toxic accumulations of chloride or boron. Salinity decreases root growth as well as shoot growth but this reduction is lower in roots compared to tops growth (Fig. 7.1). Besides generally stunting plant growth, salinity causes many specific anatomic changes that are often related to the ionic composition of the root media. Chloride

Table 7.2 Influence of salinity on P and K contents in the tops of two irrigated rice genotypes

Salinity level ($ds\ m^{-1}$ at 25 °C)	CNA 810098		CNA 810162[a]	
	P conc. ($g\ kg^{-1}$)	K conc. ($g\ kg^{-1}$)	P conc. ($g\ kg^{-1}$)	K conc. ($g\ kg^{-1}$)
0.29 (control)	2.9	35.5	2.5	34.0
5	2.7	32.5	2.0	32.8
10	2.4	25.7	2.1	24.3
15	1.5	22.5	–	–

[a] Genotype CNA 810162 did not produce dry matter at high salinity level; therefore, P and K concentrations were not determined
Source Adapted from Fageria (1985b)

salinity may cause larger epidermal cells, fewer stomata per unit leaf area, and a poorly developed xylem system. In contrast, sulfate salinity produces smaller cells and an increased number of stomata per unit leaf area.

Growth of rice plants decreases with the increase of salts in the root zone. The decrease in plant growth is related to increase in osmotic tension of soil solution which reduced absorption of water by roots due to accumulation of various ions in toxic amount (Saqib et al. 2005, 2008; Ribeiro et al. 2009). Toxic effects of sodium is highly notable effect of salinity on plant growth (Saqib et al. 2008). With the increasing concentration of Na^+ and Cl^-, concentrations of P, K^+, and Ca^{2+} reduced in the plants (Kumar et al. 2008; Fageria et al. 2011b). In the salt affected environment there is preponderance of nonessential elements over essential elements. In the salt affected soils, plants must absorb the essential nutrients from a diluted source in the presence of highly concentrated nonessential nutrients. This requires extra energy and plants sometimes unable to fulfill their nutritional requirements. There are two main stresses imposed by salinity on plant growth. One is water stress imposed by the increase in osmotic potential of the rhizosphere as a result of high salt concentration. Another stress is toxic effect of high concentration of ions. Hale and Orcutt (1987) reported that if the salt concentration is high enough to lower the water potential by 0.05–0.1 MPa then plant is under salt stress. If the salt concentration is not this high, the stress is ion stress and may be caused by one particular species of ion (Hale and Orcutt 1987).

In Table 7.2 salinity effects on absorption of P and K^+ by two rice genotypes can be observed. The concentration of two nutrients decreased with the increasing salinity level. Uptake of high amount of Na^+ and Cl^- reduced uptake of cations and anions and created nutritional disequilibrium in the plants, and reduced yield (Kumar et al. 2008; Fageria et al. 2011b). Hence, plants tolerant of salinity exclude Na^+ in absorption process and tried to maintain high concentration of K^+ in tops (Davenport et al. 2005; Saqib et al. 2005). High K^+/Na^+ ratio in plant tissue is considered a good indicator of salinity tolerance (Wei et al. 2003). High salt concentration in the root zone reduced photosynthesis of plants and increase respiration and consequently reduced growth (Khadri et al. 2006). Soil salinity is measured by concentration of salts or electrical conductivity. Effects of electrical conductivity on growth of crop species are presented in Table 7.3. The pH of

Table 7.3 Crop response to salinity influenced by electrical conductivity of saturated soil extract

Electrical conductivity (dS m^{-1} a 25 °C)	Crop response
0–2	Salinity effect is practically zero
2–4	Reduction in yield of very sensitive crops
4–8	Reduction in yield of most crops
8–16	Only tolerant crops produce satisfactory yield
>16	Few highly tolerant crops produce satisfactory

Source Adapted from Mengel et al. (2001)

saline soils is generally in the range of 7–8.5 (Mengel et al. 2001). If exchangeable sodium percentage (ESP = exchangeable Na/CEC × 100) is higher than 15 %, soils are called saline-sodic.

There are two options of reducing salinity problem. One is improving plants for adaptation to saline environment and other is improving soil conditions for good growth of plants. Second option implicates irrigation and drainage and high cost of operation. Information available in the literature indicates that beside soil recovery, use of salinity tolerant crop species or genotypes within species is a very attractive strategy for crop production in saline soils (Fageria et al. 2011b). However, a combination of these two methods may be more appropriate strategy to improve crop yields on salt affected soils.

7.2 Germplasm and Genetic Variability

Plant growth in saline soils depends on crop species, concentration of salts in the rhizosphere and environmental factors. Information about tolerance of crop species is fundamental for adequate management of salt affected soils. There is sufficient data in the literature showing relative tolerance of crop species to salinity. Data presented in Table 7.4 show salinity threshold values for different crop species, decrease in yield per unit of salinity increase beyond threshold and relative tolerance to salinity. Many crops seem to tolerate salinity equally well during seed germination and later growth stage. The salt tolerance of some crops, however, does change with growth stage. For example, barley, wheat, and corn are more sensitive to salinity during early seedling growth than during germination or later growth stages, wheat sugar beet and safflower are relatively sensitive during germination. The tolerance of soybean may either increase or decrease from germination to maturity, depending on the variety.

7.3 Stress Induction and Selection Strategy

To develop cultivars tolerant to salinity, first step is to identify genotypes tolerant to salinity. To achieve this objective evaluation of germplasm can be done under controlled conditions. The methodology for evaluation to salinity or other abiotic

Table 7.4 Salinity threshold, decrease in yield, and crop tolerance to salinity

Crop	Electrical conductivity of saturated soil extract		Classification[a]
	Threshold (dS m^{-1})	Yield decrease (% per dS m^{-1}above threshold)	
Fiber, grain, and special crops			
Cotton	7.7	5.2	T
Peanut	3.2	29.0	MS
Rice	3.0	12.0	S
Oats	–	–	MT
Sugar beet	7.0	5.9	T
Sugarcane	1.7	5.9	MS
Cowpea	4.9	12.0	MT
Rye	–	–	MT
Barley	8.0	5.0	T
Dry bean	1.0	19.0	S
Sunflower	–	–	MS
Guar	–	–	MT
Flax	1.7	12.0	MS
Millet	–	–	MS
Corn	1.7	12.0	MS
Soybean	5.0	20.0	MT
Sorghum	6.8	16.0	MT
Wheat	6.0	7.1	MT
Triticale	–	–	T
Grasses and forages			
Alfalfa	2.0	7.3	MS
Bermuda grass	6.9	6.4	T
Sudan grass	2.8	4.3	MT
Tall fescue	3.9	5.3	MT
Sesbania	2.3	7.0	MS
Ladino clover	1.5	12.0	MS
Red clover	1.5	12.0	MS
Vegetable and fruit crops			
Lettuce	1.3	13.0	MS
Asparagus	4.1	2.0	T
Potato	1.7	12.0	MS
Sweet potato	1.5	11.0	MS
Eggplant	–	–	MS
Broccoli	2.8	9.2	MS
Onion	1.2	16.0	S
Carrot	1.0	14.0	S
Cauliflower	–	–	MS
Pea	–	–	S
Spinach	2.0	7.6	MS
Watermelon	–	–	MS
Strawberry	1.0	33.0	S
Radish	0.9	9.0	MS

(continued)

Table 7.4 (continued)

Crop	Electrical conductivity of saturated soil extract		Classification[a]
	Threshold ($dS\ m^{-1}$)	Yield decrease (% per $dS\ m^{-1}$above threshold)	
Cucumber	2.5	13.0	MS
Turnip	1.2	13.0	MS
Cabbage	1.8	9.7	MS
Tomato	2.5	9.9	MS

[a] *S* Sensitive, *MS* moderately sensitive, *T* tolerant, *MT* moderately tolerant
Source Adapted from Maas (1986)

Table 7.5 Nutrient solution composition for seed germination

Reagents	Nutrient	Concentration	
		$mg\ L^{-1}$	$M \times 10^{-4}$
NH_4NO_3	N	40.00	28.170
NaH_2PO_4	P	4.00	1.290
K_2SO_4	K	40.00	10.230
$CaCl_2$	Ca	40.00	10.000
$MgSO_4.7H_2O$	Mg	40.00	16.450
$(NH_4)_6MO_7O_{24}.4H_2O$	Mo	0.05	0.005
$MnCl_2.4H_2O$	Mn	0.50	0.090
H_3BO_3	B	0.20	0.185
$ZnSO_4.7H_2O$	Zn	0.01	0.001
$CuSO_4.5H_2O$	Cu	0.01	0.001
Fe-EDTA	Fe	0.50	0.089

stress should be simple, economic, and easily adopted. Fageria (1985a) developed a technique to evaluate rice genotypes for salinity tolerance. An important component of this methodology is germinating the seeds of genotypes in solution culture. To obtained good germination, seeds should be treated with 0.1 % solution of $HgCl_2$ for 10 min and followed a washing with distilled water and put on nylon screen floating in nutrient solution. Each genotype should be germinated in a plastic pot having capacity of about two liters. Germinating pots should be covered with another bigger pot having black color or painted with black color. Composition of nutrient solution utilized in this technique is given in Table 7.5. After germination, pots covered germinating pots should be removed and seedlings should be left in the nutrient solution for about a week (Fig. 7.2). Seedlings of 7–10 days old can be transplanted in plastic tray having dimension of 30 × 45 × 8 cm with determined salinity level (Fig. 7.3). To create salinity level, 5 kg soil should be put in each tray and this soil should be mixed with a 0.5 % NaCl solution of 4 liter 3 days before transplanting. Each tray should also receive 1 g ammonium sulfate, 1.8 g triple superphosphate, 1 g potassium chloride, and 0.25 g zinc sulfate as basal fertilizer application. Twelve seedlings of each genotype should be planted in each tray. After seedling

Fig. 7.2 Growth of rice seedlings in nutrient solution

Fig. 7.3 Growth of rice cultivars at low and high salinity levels

transplanting, soil in the tray should be flooded with distilled water to a depth of about 1 cm. Four weeks after transplanting percentage of dead leaves in each tray should be counted (Table 7.6) and cultivars/genotypes should be classified for salinity tolerance. Score to classify cultivars to salinity is given in Table 7.7.

Tops dry weight can also be used as a criterion to classify cultivars tolerant to salinity. In this case, minimum two levels of salinity should be imposed. One is control and another level which can affect plant growth significantly. Equation to

Table 7.6 Classification of rice cultivars tolerance to salinity based on dead leaves

Dead leaves (%)	Score	Classification
0–20	1	Tolerant
21–35	2	Tolerant
36–50	3	Tolerant
51–70	5	Moderately tolerant
71–90	7	Moderately sensitive
91–100	9	Sensitive

Source Adapted from Ponnamperuma (1977)

Table 7.7 Classification of rice genotypes to salinity tolerance

Genotype	Dead leaves (%)	Score	Classification[a]
IR 5624-164-2-1	20	1	T
IR 9129-102-2	12	1	T
BG 11-11	18	1	T
IR 4422-164-3-6	34	2	T
IR 4432-28-5	35	2	T
Tox 711-22	35	2	T
Takatiya	38	3	T
IR22	47	3	T
IR 841-63-51-9-33	41	3	T
IR 3511-39-3-3	53	5	MT
Tox 711-16	55	5	MT
Suvale1	59	5	MT
BG 90-1	64	5	MT
De Abril	71	7	MS
IR2153-43-2-5-4	71	7	MS
Tox 711-11	71	7	MS
Labelle	77	7	MS
Paga Dívida	88	7	MS
IRGA 408	90	7	MS
IAC 435	91	9	S
BR 4	91	9	S
IAC 120	95	9	S
IR26	100	9	S
IR 8	100	9	S
EEA 405	100	9	S

[a] *T* Tolerant, *MT* moderately tolerant, *MS* moderately sensitive, *S* Sensitive
Source Fageria et al. (1981)

determine reduction of dry matter with the addition of salinity treatment and interpretation of salinity tolerance results are presented in Table 7.8. Results obtained by this criterion about rice genotypes are given in Table 7.9. Besides this criterion, dry matter or grain yield production efficiency index can also be used to classify cultivars tolerance to salinity (Fageria 1991):

Table 7.8 Classification of crop genotypes to salinity based on dry matter and grain yield reduction

Reduction in yield (%)	Classification
0–20	Tolerant
21–40	Moderately tolerant
41–60	Moderately sensitive
>60	Sensitive

$$\text{Yield reduction} = \frac{\text{Yield without salinity} - \text{Yield with salinity Yield without salinity}}{} \times 100$$

Source Fageria (1985a)

Table 7.9 Influence of salinity on dry matter yield of tops (g/5 plants) of rice genotypes and their classification for salinity tolerance

Genotype	Electrical conductivity ($dS\ m^{-1}$)			Reduction in dry matter yield (%)	
	Control (0,29)	5	10	5	10
CNA 810098	3.30	3.25	2.76	2 (T)	16 (T)
CNA 810112	3.76	2.85	0.97	24 (MT)	74 (S)
CNA 810115	4.66	3.33	1.67	29 (MT)	64 (S)
CNA 810129	2.99	2.89	1.13	3 (T)	62 (S)
CNA 810138	3.76	2.16	1.37	43 (MS)	64 (S)
CNA 810168	3.12	2.69	1.96	14 (T)	38 (MT)

T tolerant, *MT* moderately tolerant, *MS* moderately sensitive, *S* sensitive
Source Fageria (1985b, 1992)

$$\text{DM or GY efficiency index} = \frac{Y_1}{AY_1}\ x\ \frac{Y_2}{AY_2}$$

Where DM = dry matter, GY = grain yield, Y_1 = dry matter or grain yield at low salinity level, AY_1 = average dry matter or grain yield of genotypes at low salinity level, Y_2 = dry matter or grain yield at high salinity level, and AY_2 = average dry matter or grain yield of genotypes at high salinity level. When DM or GY efficiency index is higher than 1.0, cultivars should be classified as tolerant, DM or GY efficiency index in the range of 0.5–1.0, cultivars should be classified as moderately tolerant and when this index is lower than 0.5, cultivars should be classified as sensitive to salinity (Table 7.10).

7.4 Heridity, Parents Effect and Relationship Among Traits

Understanding genetic base to salinity tolerance of crop species is very important because breeding procedure for this stress depend on heredity pattern, quantitative and qualitative number of genes with major effect, and nature of their action

Table 7.10 Influence of salinity on tops dry weight of rice (g) and their classification for salinity tolerance based on yield efficiency index (YEI)

Cultivar/Lines	Salinity level (dS m^{-1})		YEI and classification[a]
	Control (0.29)	10	
GA 3459	1.16	0.42	0.60 (MT)
L 440	1.99	0.47	1.16 (T)
IET 2881	1.87	0.81	1.88 (T)
GA 3461	1.32	0.49	0.80 (MT)
CNA 12	1.92	0.56	1.33 (T)
GA 3452	1.96	0.59	1.53 (T)
CNA 294-B-BM-4-4	1.85	0.61	1.40 (T)
CNA 237-F-130-1	1.57	0.56	1.09 (T)
CNA 108-B-28-2-1	1.15	0.16	0.23 (S)
CNA 296-B-BM-M-4	1.63	0.28	0.56 (MT)
Average	1.64	0.49	

[a] *T* tolerant, *MT* moderately tolerant, *MS* moderately sensitive, *S* sensitive
Source Fageria (1985a)

(Rao and McNeilly 1999). In rice spikelet sterility which is important trait affecting yield under saline conditions at least by three genes. Diallel analysis showed effect of salinity on seedlings and spikelet sterility suggested additive as dominant and some of high heritability (Flowers 2004). There is also evidence of dominancy of salinity tolerance in sorghum. Diallel analysis, based on NaCl tolerance, expressed as root length of salt treated plant compared without salt treated plant, showed that there was an additive effect and dominant by NaCl. In corn Rao and McNeilly (1999) reported that salinity tolerance in the vegetative growth stage was governed by genes having additive effect and not the non additive. Exploiting heterosis can be helpful in finding material for salinity tolerance. Virmani (2003) reported that heterosis pattern in hybrid rice was significant in saline soil compared with normal soil condition in Egypt. In Philippines, hybrid rice presented higher tolerance to salinity compared to inbreed lines. These examples prove that salinity tolerance is genetically complex, presenting heterosis, dominant, and additive effects (Flowers 2004).

7.5 Breeding Methods

Developing crop plants tolerant to salinity is very important strategy to improve yield on salt affected soils. Ashraf et al. (2008) reported that salinity tolerance is very complex matter at both the whole plant level and the cellular level, involving interaction of stress with molecular, biochemical, and physiological processes at different stages of plant growth and development. The genetic variation to salinity is very low for most crop species which makes difficult breeding for this stress by conventional techniques. Chinnusamy et al. (2005) reported that conventional

Table 7.11 Cultivars of some crop species commercially released for cultivation

Cultivar (species)	Selection method	Country (year of release)
Arsola 1–18 (avocado)	Cultivar crossing	USA (1951)
Nebraska 10 (agropyron)	Natural selection of ecotype	USA (1962)
AZ Germ Salt 1 (alfalfa)	Back crossing selection	USA (1983)
Arizona 8601 (corn)	Natural selection program	USA (1987)
AZ Germ Salt 2 (alfalfa)	Back crossing selection	USA (1990)
Giza 159 (rice)	Cultivar crossing	Egypt (1966)
Edkway (tomato)	Natural selection program	Egypt (1982)
Giza 160 (rice)	Cultivar crossing	Egypt (1984)
Saltol (red fescue)	Natural selection of ecotype	Canada (1981)
BG 84-3 (melon)	Natural selection of ecotype	Israel (1990)

Source Adapted from Shannon (1996) and Noble and Rogers (1992)

breeding for salt tolerance is very time consuming, undesirable genes are often transferred along with desirable traits and reproductive barriers restrict transfer of favorable alleles from interspecific and intergeneric sources. However, there has been progress in developing crop cultivars tolerant to salinity by conventional breeding (Ashraf 1994). Data in Table 7.11 show commercially released cultivars of some crop species to salinity tolerance. Utilizing wild relatives of crop plants as a source of genes conferring salt tolerance can broaden the range of variation that can be used in crop production. However, Ashraf et al. (2008) reported that incorporating tolerance genes from wild relatives into domesticated crops is difficult because of reproductive barriers and there are very few examples of effectiveness of this approach in the literature.

7.6 Use of Biotechnology for Breeding to Salinity Tolerance

Conventional breeding had contributed to developing crop cultivars for salinity tolerance. However, the progress so far has not been significant. Hence, modern techniques like soma-clonal variation, protoplasmic fusion, and mutation breeding can contribute significantly in breeding crop cultivars for salinity tolerance, if the traits for salt tolerance do not exist or if the genetic variability for specific traits is absent (Ashraf 1994). The prime strategy in mutation based breeding is to induce or alter one or two major traits through mutagens, which may be chemical or radiation (Ashraf et al. 2008). There has been experimental evidence that rice mutants NIAB Rice- 1 and PSR 1–84 have shown greater yield than their respective salt-tolerant check cultivars, Pokkali and Johna 349, under saline conditions (NIAB 1987). In addition, the use of chemical mutagens in developing salt-tolerant rice mutants has also been shown to be successful. Ashraf (1984) induced variability for salt tolerance in the salt sensitive rice cultivar Taichung 65 by treating fertilizer egg cells with varying doses of N-methyl-N-nitrosourea.

In M_3, two salt-tolerant mutants were detected that had 83 % and 90 % survival at the seedling stage in 0.5 % NaCl.

Genetic markers (RAPD, AFLP, and SSR) that have been routinely used in fingerprints, genetic mapping, and quantitative trait loci (QLT) analysis may help in identifying useful mutations (Mlcochová et al. 2004) induced by chemical mutagens or radiations and may help in examining its physiological bases, which will bring a new dimension in gene technology. Molecular biology can also play a significant role in incorporating salt resistant genes in crop plants. Advance in molecular biology leads to development of molecular marker of DNA which can be used to identify QTLs. Over the last two decades, advances in molecular marker technology have led to the development of detailed molecular linkage maps for many plant species. The DNA based markers speed up the advance in improvement of crops for stress tolerance (Vinh and Paterson 2005). Furthermore, DNA markers can be used as a diagnostic tool to identify genotypes in large populations that bring together superior genetic potential for productivity under stress, with the traits that are necessary to develop a potential commercial cultivar (Vinh and Paterson 2005; Ashraf et al. 2008).

The effectiveness of QTL mapping in transferring stress genes is very high. It is reported that QTL mapping is a way to estimate the locations, numbers, extent of phenotypic effects and modes of gene action, and of individual determinants that substantially contribute to the inheritance of continuously variable traits (Vinh and Paterson 2005). The principal objective is to isolate genetic signals emerging from an individual locus, the background collective effects of nongenetic factors, as well as measurement errors in evaluation of continuous traits. The QTL mapping is thus an effective means for identifying specific components that allow direct assessment of stress tolerance (Ashraf et al. 2008). The QTLs and marker-assisted selection offer several advantages over direct phenotypic screening, in as much as the PCR-based techniques used to identify the markers reduce the time needed to screen genotypes as well as reduce the environmental impact on the trait under study.

There are several reports in the literature that genetic engineering is an important tool in gene transfer across reproductive barriers. In addition, genetic engineering offers a way to create new genetic variation when natural allelic variation may be limited or inappropriate (Humphreys and Humphreys 2005). Yamaguchi and Blumwald (2005) reported that there are two genetic approaches which are currently being used to improve crop stress tolerance These approaches are: (1) utilization of natural genetic variation, either through direct selection in stressful conditions or through the mapping of QTLs and subsequent marker-assisted selection, and (2) production of transgenic plants by introducing the novel genes or by modifying the expression of the existing genes to alter degree of stress tolerance. Large-scale screening of crop genotypes for salinity tolerance can be done by dissection of complex salt tolerance trait by means of QTL mapping and identifying chromosomal regions associated with DNA markers (Ashraf et al. 2008).

7.7 Transgenic Method

Ashraf et al. (2008) defined transgenic crops are bioengineered crops that possess a gene or genes that have been inserted by human into their genome using modern biotechnology. These authors further said that the inserted gene sequence, known as the transgene, may belong to an unrelated plant, or even to a bacterium or animal. Crops that contain these transgene are also described as genetically modified and are often referred to as genetically modified plants (Ashraf et al. 2008). Transgenic approach is widely used by plant scientists to develop crop cultivars to biotic and abiotic stresses. Advances in molecular biology have led the identification of a large number of genes that are induced as a result of drought or salinity stress (Lea et al. 2004). The major focus of research suing transgenic method or approach are genes that encode: (i) compatible osmotic (GB, proline, sugars) (ii) transcription factors (iii) enzymatic and nonenzymatic antioxidants (iv) ion transport proteins, and (v) heat shock and late embryogenesis abundant (LEA) proteins (Ashraf et al. 2008).

In transgenic rice plants, overexpressing peroxisomal BADH exhibited enhanced ion selectivity under saline conditions by accumulating high amounts of K^+ but low amounts of Na^+ and Cl^- (Kishitani et al. 2000). However, it is not yet clear how GB overproduction increases the K^+/Na^+ ratio, which is regulated by ion transporters and channels (Ashraf et al. 2008). In general, proline accumulation in response to abiotic stresses including salt stress is found to be correlated with stress tolerance of many plants (Ashraf and Harris 2004). Zhu et al. (1998) reported that genetically modified rice plants showed faster recovery after a short period of salt stress compared to inbred rice plants.

Although genetic engineering plants to overproduce proline or GB seems to be an effective means of increasing salinity tolerance, some failures of this approach are also reported in the literature (Ashraf and Foolad 2007). It is also apparent that salt tolerance in plants depends on both the level of accumulation of compatible solutes and their subsequent transportation to target compartments (Chinnusamy et al. 2005). For example, transgenic plants expressing choline oxidase targeted to chloroplasts exhibited higher resistance to photoinhibition under salt and cold stresses than did transgenic plants with choline oxidase targeted to the cytosol (Sakamoto et al. 1998). Considering various strategy available to induce salt tolerance in plants, it is suggested that all factors associated with gene regulation at transcriptional and translational levels should be examined while engineering plants for these compatible solutes (Ashraf et al. 2008).

Ashraf et al. (2008) and Garg et al. (2002) reported that transgenic rice expressing the trehalose gene is responsible for multi stress tolerance such as salinity, drought, and cold stress. Similarly, transgenic rice plants expressing chimeric gene Ubi1::TPSP accumulated a high levels of trehalose that resulted in increased tolerance of drought, salt, and cold, as shown by chlorophyll florescence and growth inhibition analyses (Jang et al. 2003). However, several pleiotropic effects observed in these transgenic led to the suggestion that trehalose affects other plant development process as well (Ashraf et al. 2008).

Sugar alcohols, such as glycerol, mannitol, sorbitol, and D-ononitol, are potential osmoprotectants in many halophytes (Yancey et al. 1982). Transgenic tobacco and wheat plants expressing the *mtlD* gene, responsible for the biosynthesis of mannitol, were found to be tolerant to salt stress (Tarczynski et al. 1993, Abebe et al. 2003).

Ectoine is another important compatible solute and is generally accumulated in halophytic bacteria. Its introduction into plants can also alleviate adverse effects of salt stress. Tobacco plants transformed with three genes isolated from halophytic bacteria *Halomas elongata*, which are responsible for biosynthesis of ectoine, when treated with NaCl accumulated more ectoine and showed greater growth than untransformed plants (Moghaieb et al. 2006).

Plants can produce various antioxidants and detoxifying enzymes to efficiently scavenge reactive oxygen species (ROS), that can cause considerable oxidative damage to membrane lipids, proteins, and nucleic acids. The various antioxidants used by plants are ascorbate, glutathione, α-tocopherol, and carotenoids, whereas detoxifying enzymes include superoxide dismutase (SOD), catalase, peroxidase, and enzymes of ascorbate–glutathione cycle. However, it is crucial to target enzymes at the site where the stress-induced ROS production takes place for detoxification and hence improved stress tolerance. This indicates that proteins that are damage by oxidative stress have a significant adverse effect on plant tolerance to environmental stresses. Tobacco plants modified to overexpress glutathione S-transferase/glutathione peroxidase, one of the major enzymes of ascorbate–glutathione cycle, were found to be tolerant to both chilling and salt stresses (Roxas et al. 1997, 2000).

Most oxidative stress is triggered by methionine oxidation resulting in disruption of protein structure (Hoshi and Heinemann 2001). Oxidized methionine can be reduced back to methionine by the activity the enzyme methionine sulfoxide reductase (MSR) (Sadanandom et al. 1996). Romero et al. (2004) demonstrated that plant lines overexpressing MSR4 in chloroplasts have increased resistance to oxidative damage and are expected to be salt tolerant because higher antioxidant capacity of a plant is associated with salt tolerance (Gossett et al. 1996).

Some aldehyde dehydrogenases are known to have a role in osmoregulation by catalyzing the synthesis of osmoprotectants (Kirch et al. 2004). Rodrigues et al. (2006) showed that both tobacco and *Arabidopsis* expressing aldehyde dehydrogenase *GmALDH7* showed greater germination and reduced reactive aldehydes, which are generated by lipid peroxidation, under saline conditions. These findings indicate that *GmALDH7* is one of the most effective genes for producing salt-stress-tolerant plants.

Aharon et al. (2003) found that overexpression of the vacuolar Na^+/H^+ antiporter that sequesters Na^+ in vacuoles (NHX1) improved the salinity tolerance in *Arabidopsis*, tomato, and *Brassicas*. Ohta et al. (2002) and Fukuda et al. (2004) found that the salt tolerance of transgenic rice overexpressing halophyte (*Atriplex gmelini*) gene *AgNHX1* and rice gene *OsNHX1* was improved compared with wild types. However, the increase of leaf Na^+ was similar in both the transgenic and wild-type plants. These results indicate that the Na^+/H^+ antiporter gene could sequester part of the Na^+ in the vacuoles and prevent the toxic effects of excessive Na^+ ions on the cells.

Active transport of Na^+ across plant cell membranes is usually coupled to the proton (H^+) electrochemical potential established by H^+-translocating pumps (Gaxiola et al. 2001). From this finding, it is suggested that overexpression of cation transporters in combination with H^+-translocating pumps can increase salt tolerance. Zhao et al. (2006) reported that the coexpression of *Suaeda salsala SsNHX1* and *Arabidopsis AVP1* in transgenic rice caused greater improvement in salt tolerance than transformation with the single gene, *SsNHX1*.

During the last two decades, many abiotic stress-inducible genes have been cloned and characterized from different plant species. However, for the expression of these genes, there is a need to identify suitable promoters. Efforts have been made to identify and characterize stress-induced promoters, particularly those induced by anaerobic conditions, low or high temperatures or salt stress (Grover et al. 2001). For example, the production of transgenic plants with *DREB* genes is useful for improvement of tolerance of environmental stresses in a number of species. However, constitutive expression of these genes retards plant growth. Development of transgenic plants with stress-inducible promoters along with *DREB* genes or regulation of expression of *DREB* genes by stress-inducible promoters can induce stress tolerance and minimize the adverse effects of stress on growth. For the effective application of molecular approaches to producing stress-tolerant plants, it would be desirable to restrict transgene expression to particular tissues by the use of tissue-specific promoters (Gittins et al. 2001).

Several groups of late embryogenesis abundant (LEA) protein genes have been demonstrated to confer water deficit and salt-stress tolerance. Expression of *HVA1*, a group 3 LEA protein from barley, conferred tolerance to soil water deficit and salt stress in transgenic rice plants (Babu et al. 2004).

Although transgenic approaches for enhance abiotic stress tolerance are gaining ground among both scientists and public, the achievements made so far are not significant. This has been due to the fact that scientists have been producing transgenic of various crops in the past using single-gene transfer that undoubtedly resulted in transgenics with limited stress tolerance. However, there is a growing trend now to use the multigene approach by which several genes responsible for overall stress tolerance are simultaneously transferred to the transgenics (Cherian et al. 2006). Furthermore, other protocols such as RNAi and transposon insertional knockouts for the candidate stress-tolerant genes and signaling pathways show a great promise to produce highly stress-tolerant crop plants.

7.8 Conclusions

Salinity is a serious problem in worldwide including Brazil. Salt affected soils can be defined as those soils that have been adversely modified for the growth of most crop plants by the presence of soluble salts, with or without high amounts of exchangeable sodium. Civilizations have been destroyed by the encroachment of salinity on the soils, as a result of vast areas of the land is rendered unfit for agriculture. Salt affected soils are found in many regions of the world. Salt affected soils normally occur in arid and semi-arid regions where rainfall is insufficient to leach salts from the root zone. Salt problems, however, are not restricted to arid or semi-arid regions. They can develop even in sub-humid and humid regions under appropriate conditions. In addition, these soils may also occur in coastal areas subject to tides. Salts generally originate from native soil and irrigation water. Roughly 263 million hectares are irrigated area worldwide and in most of that area salinity is a growing threat. The irrigated area represents about 20 % of the total land used for crop production. This represents about 19 % of the total area of the world under crop production. Use of inappropriate levels of fertilizers with inadequate management practices can create saline conditions even in humid conditions. Salinization is the process whereby soluble salts accumulate in the root zone. Common ions contributing to this problem are Ca^{2+}, Mg^{2+}, Cl^-, Na^+, SO_4^{2-}, HCO_3^- and in some cases K^+ and NO_3^-. Salinity can be measured by determining electrical conductivity of the saturated soil extract.

The important soils and plant management practices which can improve crop yield on salt affected soils are use of soil amendments to reduce effect of salts, application of farmyard manures to create favorable plant growth environments, leaching salts from soil profile, and planting salt-tolerant crop species or genotypes within species. Addition of fertilizers, especially potassium may also help in reducing salinity effects and improving nutrient use efficiency. Breeding salt tolerance cultivars by conventional methods has limited success due to several reasons. Hence, genetic engineering seems to offer considerable promise for the development of salinity tolerant plants, including cultivars of important food crops.

References

Abebe T, Guenzi AC, Martin B, Cushman JC (2003) Tolerance of mannitol-accumulating transgenic wheat to water stress and salinity. Plant Physiol 131:1748–1755

Aharon GS, Apse MP, Duan S, Hua X, Blumwald E (2003) Characterization of a family of vacuolar Na^+/H^+antiporters in *Arabidopsis thaliana*. Plant Soil 253:245–256

Ashraf M (1984) Induced variability for salinity and aluminium tolerance by N-Methyl N-Nitrosourea treatment of fertilized egg cell of rice. Dissertation, University of Philippines

Ashraf M (1994) Breeding for salinity tolerance in plants. Crit Rev Plant Sci 13:17–42

Ashraf M, Athar HR, Harris PJC, Kwon TR (2008) Some prospective strategies for improving crop salt tolerance. Adv Agron 97:45–110

Ashraf M, Foolad MR (2007) Roles of glycine betaine and proline in improving plant abiotic-stress resistance. Environ Exper Bot 59:206–216

Ashraf M, Harris PJC (2004) Potential biochemical indicators of salinity tolerance in plants. Plant Sci 166:3–16

Babu RC, Zhang J, Blum A, Ho DHT, Wu DR, Nguyen HT (2004) HVA1, a LEA gene from barley confers dehydration tolerance in transgenic rice (*Oryza sativa* L.) via cell membrane protection. Plant Sci 166:855–862

Cherian S, Reddy MP, Ferreira RB (2006) Transgenic plants with improved dehydration-stress tolerance: Progress and future prospects. Biol Plantarum 50:481–495

Chinnusamy V, Jagendorf A, Zhu JK (2005) Understanding and improving salt tolerance in plants. Crop Sci 45:437–448

Davenport R, James RA, Zakrisson-Plogander A, Tester M, Munns R (2005) Control of sodium transport in durum wheat. Plant Physiol 137:807–818

Epstein E, Bloom AJ (2005) Mineral nutrition of plants: principles and perspectives. Sinauer, Sunderland

Fageria NK (1985a) Relatório do projeto "Avaliação de cultivares de arroz para condições adversas do solo". Embrapa-CNPAF, Goiânia

Fageria NK (1985b) Salt tolerance of rice cultivars. Plant Soil 88:237–243

Fageria NK (1991) Tolerância de cultivares de arroz à salinidade. Pesq Agropecu Bras 26: 281–288

Fageria NK (1992) Maximizing crop yields. Dekker, New York

Fageria NK, Baligar VC, Jones CA (2011a) Growth and mineral nutrition of field crops. CRC Press, Boca Raton

Fageria NK, Barbosa Filho MP, Gheyi HR (1981) Avaliação de cultivares de arroz para tolerância à salinidade. Pesq Agropecu Bras 16:677–681

Fageria NK, Gheyi HR, Moreira A (2011b) Nutrient bioavailability in salt affected soils. J Plant Nutr 34:945–962

Fageria NK, Soares Filho WS, Gheyi HR (2010) Melhoramento genético vegetal e seleção de cultivares tolerantes à salinidade. In: Gheyi HR, Dias NS, Lacerda CF (eds) Manejo da salinidade na agricultura: Estudos básicos e aplicados. INCT Sal, Fortaleza

Flowers TJ (2004) Improving crop salt tolerance. J Exper Bot 55:307–319

Fukuda A, Nakamura A, Tagiri A, Tanaka H, Miyao A, Hirochika H, Tanaka Y (2004) Function, intracellular localization and the importance in salt tolerance of a vacuolar Na^+/H^+antiporter from rice. Plant Cell Physiol 45:146–159

Garg AK, Kim JK, Owens TG, Ranwala AP, Do Choi Y, Kochian LV, Wu RJ (2002) Trehalose accumulation in rice plants confers high tolerance levels to different abiotic stresses. Proc Natl Acad Sci USA 99:15898–15903

Gaxiola RA, Li J, Undurraga S, Dang LM, Allen GJ, Alper SL, Fink GR (2001) Drought- and salt-tolerant plants result from overexpression of the AVP1 H^+-pump. Proc Natl Acad Sci USA 98:11444–11449

Gittins JR, Hiles ER, Pellny TK, Biricolti S, James DJ (2001) The *Brassica napus* extA promoter: A novel alternative promoter to CaMV 35S for directing transgene expression to young stem tissues and load bearing regions of transgenic apple trees (*Malus pumila* Mill.) Mol Breed 7:51–62

Gossett DR, Banks SW, Millhollon EP, Lucas MC (1996) Antioxidant response to NaCl stress in a control and a NaCl-tolerant cotton line grown in the presence of paraquat, buthioninesulfoxime, and exogenous glutathione. Plant Physiol 112:803–809

Grover A, Kapoor A, Lakshmi OS, Agarwal S, Sahi C, Agarwal SK, Agarwal M, Dubey H (2001) Understanding molecular alphabets of the plant abiotic stress responses. Curr Sci 80:206–216

Hale MG, Orcutt DM (1987) The physiology of plants under stress. Wiley, New York

Hoshi T, Heinemann SH (2001) Regulation of cell function by methionine oxidation reduction. J Physiol 531:1–11

Humphreys MO, Humphreys MW (2005) Breeding for stress resistance: general principles. In: Ashraf M, Harris PJC (eds) Abiotic stresses: plant resistance through breeding and molecular approaches. Haworth Press, New York

Jang IC, Oh SJ, Seo JS, Choi WB, Song SI, Kim CH, Kim YS, Seo HS, Do Choi Y, Nahm BH, Kim JK (2003) Expression of a bifunctional fusion of the *Escherichia coli* genes for trehalose-6-phosphate synthase and trehalose-6-phosphate phosphatase in transgenic rice plants increase trehalose accumulation and abiotic stress tolerance without stunting growth. Plant Physiol 131:516–524

Khadri M, Tejera NA, Lluch C (2006) Sodium chloride-ABA interaction in two common bean (*Phaseolus vulgaris*) cultivars differing in salinity tolerance. Environ Exper Bot 60:211–218

Kirch HH, Bartels D, Wei Y, Schnable PS, Wood AJ (2004) The ALDH gene super family of *Arabidopsis*. Trends Plant Sci 9:371–377

Kishitani S, Takanami T, Suzuki M, Oikawa M, Yokoi S, Ishitani M, Alvarez-Nakase AM, Takabe T, Takabe T (2000) Compatibility of glycinebetaine in rice plants: evaluation using transgenic rice plants with a gene for peroxisomalbetaine aldehyde dehydrogenase from barley. Plant, Cell Environ 23:107–114

Kumar V, Shriram V, Nikam TD, Jawali N, Shitole MG (2008) Sodium chloride-induced changes in mineral nutrients and proline accumulation in indica rice cultivars differing in salt tolerance. J Plant Nutr 31:1999–2017

Lal R, Hall GF, Miller FP (1989) Soil degradation. I. basic processes. Land Degrad Rehabil 1:51–69

Lea PJ, Parry MAJ, Medrano H (2004) Improving resistance to drought and salinity in plants. Ann Appl Biol 144:249–250

Maas EV (1986) Salt tolerance of plants. Appl Agric Res 1:12–25

Mengel K, Kirkby EA, Kosegarten H, Appel T (2001) Principles of plant nutrition. Kluwer, Dordrecht

Mlcochová L, Chloupek O, Uptmoor R, Ordon F, Friedt W (2004) Molecular analysis of the barley cv. 'Valticky' and its X-ray-derived semidwarf-mutant 'Diamant'. Plant Breed 123:421–427

Moghaieb REA, Tanaka N, Saneoka H, Murooka Y, Ono H, Morikawa H, Nakamura A, Nguyen NT, Suwa R, Fujita K (2006) Characterization of salt tolerance in ectoine-transformed tobacco plants (*Nicotiana tabaccum*): Photosynthesis, osmotic adjustment, and nitrogen partitioning. Plant Cell Environ 29:173–182

National Academy of Sciences (1999) Water for the future: the West Bank Gaza strip, Israel, and Jordan. National Academy Press, Washington

NIAB (1987). Nuclear Institute for Agriculture and Biology. Fifteen years of NIAB. NIAB, Faisalabad

Noble CL, Rogers ME (1992) Arguments for the use of physiological criteria for improving the salt tolerance in crops. Plant Soil 146:99–107

Ohta M, Hayashi Y, Nakashima A, Hamada A, Tanaka A, Nakamura T, Hayakawa T (2002) Introduction of a Na^+/H^+antiport gene from Atriplex gmelini confers salt tolerance to rice. FEBS Lett 532:279–282

Pessarakli M, Szabolcs I (1999) Soil salinity and sodicity as particular plant/crop stress factors. In: Pessarakli M (ed) Handbook of plant and crop stress. Dekker, New York

Ponnamperuma FN (1977) Screening rice for tolerance to mineral stress. IRRI, Los Baños

Rains DW, Goyal SS (2003) Strategies for managing crop production in saline environments: An overview. In: Sharma SK, Rains DW, Morgan RL (eds) Goyal SS. Haworth, New York

Rao SA, McNeilly T (1999) Genetic basis of variation for salt tolerance in maize (*Zea mays* L.). Euphytica 108:145–150

Ribeiro MS, Barros MFC, Santos MBG (2009) Química dos solos salinos e sódicos. In: Melo VF, Alleoni LRF (eds) Química e mineralogia do solo. Parte II-Aplicações. Sociedade Brasileira de Ciência do Solo, Viçosa

Rodrigues SM, Andrade MO, Gomes APS, Da Mata FM, Baracat-Pereira MC, Fontes EPB (2006) Arabidopsis and tobacco plants ectopically expressing the soybean antiquitin-like ALDH7 gene display enhanced tolerance to drought, salinity, and oxidative stress. J Exper Bot 57:1909–1918

Romero HM, Berlett BS, Jensen PJ, Pell EJ, Tien M (2004) Investigations into the role of the plastidal peptide methionine sulfoxidereductase in response to oxidative stress in *Arabidopsis*. Plant Physiol 136:3784–3794

Roxas VP, Lodhi SA, Garret DK, Mahan JR, Allen RD (2000) Stress tolerance in transgenic tobacco seedlings that overexpress glutathione S-transferase/glutathione peroxidase. Plant Cell Physiol 41:1229–1234

Roxas VP, Smith RK Jr, Allen ER, Allen RD (1997) Overexpression of glutathione S-transferase/ glutathione peroxidase enhances the growth of transgenic tobacco seedlings during stress. Nat Biotechnol 15:988–991

Sadanandom A, Piffanelli P, Knott R, Robinson C, Sharpe A, Lydiate D, Murphy DJ, Fairbairn DJ (1996) Identification of a peptide methionine sulfoxidereductase gene in an oleosin promoter from *Brassica napus*. Plant J 10:235–242
Sakamoto A, Murata A, Murata N (1998) Metabolic engineering of rice leading to biosynthesis of glycinebetaine and tolerance to salt and cold. Plant Mol Biol 38:1011–1019
Saqib M, AKhtar J, Qureshi RH (2008) Sodicity intensifies the effect of salinity on grain yield and yield components of wheat. J Plant Nutr 31:689–701
Saqib M, Zorb C, Rengel Z, Schubert S (2005) Na + exclusion and salt resistance of wheat (*Triticumaestivum*) are improved by the expression of endogenous vacuolar Na^+/H^+ antiporters in roots and shoots. Plant Sci 169:959–965
Shannon MC (1996) New insights in plant breeding efforts for improved salt tolerance. HortTech 6:96–98
Soil Science Society of America (2008) Glossary of soil science terms. American Society of Soil Science, Madison
Szabolcs I (1994) Soil salinization. In: Pessarakli M (ed) Handbook of plant crop stress. Dekker, New York
Tarczynski MC, Jensen RG, Bohnert HJ (1993) Stress protection of transgenic tobacco by production of the osmolyte mannitol. Science 259:508–510
Vinh NT, Paterson AH (2005) Genome mapping and its implications for improving stress resistance in plants. In: Ashraf M, Harris PJC (eds) Abiotic stresses: Plant resistance through breeding and molecular approaches. Haworth, New York
Virmani SS (2003) Advances in hybrid rice research and development in the tropics. In: Virmani SS, Mao CX, Hardy B (eds) Hybrid rice for food security, poverty alleviation, and environmental protection. IRRI, Los Baños
Wei W, Bilsborrow PE, Hooley P, Fincham DA, Lombi E, Forster BP (2003) Salinity induced differences in growth, ionic distribution, and partitioning in barley between the cultivar Maythorpe and its derived mutant Golden Promise. Plant Soil 250:183–191
Yamaguchi T, Blumwald E (2005) Developing salt-tolerant crop plants: challenges and opportunities. Trends Plant Sci 10:615–620
Yancey PH, Clark ME, Hand SC, Bowles RD, Somero GN (1982) Living with water stress: evolution of osmolyte system. Sci 217:1214–1222
Zhao F, Zhang X, Li P, Zhao Y, Zhang H (2006) Co-expression of the *Suaeda salsa* SsNHX1 and *Arabidopsis* AVP1 confer greater salt tolerance to transgenic rice than the single SsNHX1. Mol Breed 17:341–353
Zhu B, Su J, Chang MC, Verma DPS, Fan YL, Wu R (1998) Overexpression of a pyrroline-5-carboxylate synthetase gene and analysis of tolerance to water and salt stress in transgenic rice. Plant Sci 139:41–48

Chapter 8
Breeding for Aluminum Tolerance

Lauro José Moreira Guimarães, Claudia Teixeira Guimarães, Jurandir Vieira Magalhães, Sidney Netto Parentoni and Flávia Ferreira Mendes

Abstract Aluminum (Al) toxicity is one of the greatest limitations to agriculture in acid soils, principally in the tropical regions. In these areas, crop management practices, including massive application of limestone, were adopted for the implementation of commercial agriculture. However, even in well-managed soils, subsurface layers might have a high acidity and high Al saturation, limiting root development in sensitive cultivars. Thus, breeding for Al tolerance is a key point to achieve high yield and to enhance the stability of cultivars. Nowadays, great effort has been applied for identification of genes associated to the Al tolerance aiming the incorporation of favorable alleles in elite cultivars through the integration of conventional breeding and molecular tools. In this chapter, we are discussing some topics related to the effects of Al in plants, mechanisms and inheritance of Al tolerance, passing to practical points like evaluation of this trait in annual crops, strategies and breeding methods, and, finally, biotechnology applied to the breeding for tolerance to Al toxicity.

Keywords Abiotic stress · Acid soils · Aluminum toxicity

L. J. M. Guimarães (✉) · C. T. Guimarães · J. V. Magalhães · S. N. Parentoni
Embrapa Maize and Sorghum, Sete Lagoas-MG, Brazil
e-mail: lauro@cnpms.embrapa.br

C. T. Guimarães
e-mail: claudia@cnpms.embrapa.br

J. V. Magalhães
e-mail: jurandir@cnpms.embrapa.br

S. N. Parentoni
e-mail: sidney@cnpms.embrapa.br

F. F. Mendes
Universidade Federal de Lavras, Lavras-MG, Brazil
e-mail: flvmendes2001@yahoo.com.br

R. Fritsche-Neto and A. Borém (eds.), *Plant Breeding for Abiotic Stress Tolerance*,
DOI: 10.1007/978-3-642-30553-5_8,

8.1 Introduction

Aluminum (Al) toxicity is one of the greatest limitations to agricultural production in acid soils. It is estimated that approximately 50 % of the cultivable soils in the world possess high acidity, principally in the tropical regions (Sanchez and Salinas 1981). Some soils in temperate regions with intense agricultural activity are also passing through processes of acidification as a result of the massive application of nitrogen-containing fertilizers, deforestation, and the removal of plant cover.

Acid soils represent about 70 % of Brazilian territory (Adámoli et al. 1985), including the Cerrado region, which occupies more than two million square kilometers. In this region, oxisols, which have good physical characteristics, deep profile, good water-holding capacity, and topography favorable to mechanization, predominate. However, the high saturation of exchangeable aluminum, combined with high acidity, low availability of phosphorus (P), and low cation exchange capacity (CEC), are the main limitations (Buol 2009).

Crop management practices were adopted for the implementation of commercial agriculture in the Brazilian Cerrado region. These practices include the massive application of limestone, gypsum, and phosphate fertilizers, in addition to deployment of no-tillage system using crop rotation and soil cover crops such as millet and signal grass. Also important was the adoption of cultivars adapted to regions initially considered marginal to the production process.

Despite advances in the management of Cerrado soils, the correction of soil acidity to neutralize the toxic Al in layers 20 cm below the surface is costly and difficult to operationalize. Thus, even in well-managed soils, subsurface layers might have a high acidity and high degree of saturation with toxic forms of Al, limiting root development in a cultivar sensitive to this element.

Thus, breeding for tolerance to Al toxicity is a key point to achieving high levels of yield and yield stability in environments where Al is a limiting factor. The identification of new sources of tolerance to Al toxicity and incorporation of favorable alleles through the integration of conventional breeding and molecular tools are important for reaching this objective.

8.1.1 Effect of Aluminum Toxicity in Plants

In acid soils, with pH lower than 5.0, Al is dissolved from compounds such as $Al(OH)_3$ and Al-silicates causing toxic effects in plants (Martin 1992). Al causes a rapid inhibition of root growth because of changes in cell division and elongation in the root apex (Kochian et al. 2004). Then, an Al sensitive cultivar will produce a poorly developed root system (Fig. 8.1), unable to explore deeper layers of the soil, limiting water and nutrients uptake. As a result, cultivars sensitive to Al toxicity are also susceptible to water stress and nutritional deficiency when grown in acid soils, resulting in low yield and stability.

Al toxicity can cause root abnormalities such as decrease in the number of secondary roots and root hairs, as well as thickening and stiffening of the roots,

Fig. 8.1 Development of the root system of maize plants grown in soils with different levels of Al saturation (0, 36, and 60 %). On the *left*, a sensitive genotype and, on the *right*, a genotype tolerant to Al. Photo provided by Sidney Netto Parentoni

making them more brittle. Additionally, changes can occur in physiological and biochemical processes such as phosphorylation of sugars, respiration, photosynthesis, activity of ATPases and acid phosphatases, amino acid synthesis, and permeability of the plasma membrane (Ferreira et al. 2006).

8.1.2 Mechanisms of Al Tolerance

Plants have developed various mechanisms to overcome the effects of toxic Al in the soil. These mechanisms can be divided into two groups: (i) symplastic mechanisms involving immobilization or neutralization of Al within the cell, and (ii) exclusion or apoplastic mechanisms that prevent the Al from penetrating into the cell, by its immobilization or neutralization in the rhizosphere (Samac and Tesfaye 2003; Kochian 1995).

In the symplastic mechanism, Al within the cell can form complexes with organic acids (Foy 1988; Taylor 1988), with proteins, or other compounds (Suhayda and Haug 1985). Internal Al is kept inactive in the cytoplasm or in the vacuoles, preventing its negative effects in many cellular processes. However, the intracellular mechanisms of tolerance are not well understood, since both tolerant and sensitive plants have an accumulation of Al when grown under conditions of high availability of this element. Different forms of Al can be transported into vacuoles, where it is stored without causing further damage to the cell.

The mechanisms of Al exclusion are the most well studied (Kochian et al. 2004; Samac and Tesfaye 2003) and validated on the basis of genetic, physiological, and molecular evidences. In these mechanisms, chelating compounds are released by the roots, forming non-toxic compounds with Al, avoiding the entry of this element into cells. The exudation of organic acids by root apices is a major mechanism of Al tolerance in several species, including maize (Piñeros et al. 2002), wheat (Sasaki et al. 2004), and sorghum (Magalhães et al. 2007). Organic acids, especially citrate and malate, form stable complexes with the Al^{3+} in the rhizosphere, reducing the toxic effects in the root system (Kochian et al. 2004).

The first gene for Al tolerance isolated in plants was the *ALMT1* gene in wheat, a malate transporter activated by Al (Sasaki et al. 2004). The *SbMATE* (Magalhães et al. 2007) and *HvMATE* (Furukawa et al. 2007) genes, respectively, were further isolated from sorghum and barley, functioning as citrate transporters, also induced by Al. Homologous genes of the ALMT and MATE multigene families were subsequently isolated from several other plant species. In addition, a transcription factor of the zinc finger type, called *STOP1*, is related to Al tolerance in *Arabidopsis*, regulating the expression of *AtMATE* and *AtALMT* genes (Liu et al. 2009; Iuchi et al. 2007). Gene families belonging to ABC transporters have also been associated with Al tolerance by mechanisms not yet fully elucidated (Huang et al. 2009).

Other mechanisms of exclusion involving exudation of phenolic compounds (Ofei-Manu et al. 2001), formation of mucilage (Miyasaka and Hawes 2001), and increasing of rhizosphere pH (Degenhardt et al. 1998) are also involved with Al tolerance in plants. Recently, the metal transporter *Nramp aluminum transporter 1* (*Nrat1*), expressed in the plasma membrane and in the tonoplast, was associated with Al tolerance in rice (Xia et al. 2010), suggesting a possible involvement with the influx of Al and its mobilization to the vacuole.

8.2 Germplasm and Genetic Variability

Different plant species present different behaviors in soils with high Al saturation. Parentoni et al. 2001 presented some species considered to be extremely tolerant to Al: *Andropogon gayanus*, *Brachiaria decumbens*, cassava (*Manihot esculenta* Crantz), *Stylosanthes capitata*, *Stylosanthes guianensis*, *Hyparrhenia rufa,* and cowpea (*Vigna unguiculata*). Species such as rice, coffee, potato, rubber, palm oil, rye, and oat are considered to be highly tolerant to Al. However, sweet potato, maize, cabbage, wheat, millet, pea, eggplant, soybean, elephant grass, barley, onions, beet, pumpkin, sorghum, and leucaena are crops that present low to medium tolerance to aluminum. Carrots, spinach, celery, cotton, common bean, and alfalfa are among the species that are extremely sensitive to Al.

In addition to interspecific variability, a large degree of intraspecific variability in Al tolerance has been observed in several crop species (Samac and Tesfaye 2003; Parentoni et al. 2001). A wide phenotypic variability for Al tolerance was observed in a group of 13 sorghum lines, ranging from highly sensitive (20 % relative root growth) to highly tolerant (>100 % relative root growth), when measured in nutrient solution containing 60 μM of Al activity (Caniato et al. 2007). Similarly, a wide variability in Al tolerance in nutrient solution containing toxic levels of Al has been observed in tropical maize lines. Therefore, the identification, selection, and recombination of sources of alleles for Al tolerance are essential for obtaining genetic gains in breeding programs.

8.3 Inheritance of Tolerance Al Toxicity

The genetic control of Al tolerance in crops can vary from an inheritance controlled by one or two genes, as observed in wheat, to a quantitative inheritance, where genes with smaller effects act as modifiers, such as in maize (Ferreira et al. 2006; Cançado et al. 2002). Al tolerance has been widely studied in grasses leading to discovery of genes that corroborated with the physiological mechanisms proposed so far. In wheat, the tolerance to Al appears to be controlled by one or two major genes, with the main gene located on chromosome 4D (Lagos et al. 1991; Aniol and Gustafson 1984). Delhaize et al. (1993) associated the *Alt1* locus with a large proportion of the variability in tolerance among wheat cultivars. Subsequently, the *ALMT1* gene, which encodes a malate transporter activated by aluminum, was cloned by Sasaki et al. (2004) and would be the gene underlying the *Alt1* locus.

Simple inheritance of Al tolerance was also observed in barley (Minella and Sorrells 1992; Stølen and Andersen 1978). Minella and Sorrells 1992, 1997 identified a gene (*Alp*) in this species that had a major effect in Al tolerance. They concluded that the variations in Al tolerance among barley cultivars were controlled by different alleles at this locus; however, other genes with smaller effects may have influence on this trait. The *Alp* gene was further mapped to the chromosome 4H (Tang et al. 2000).

Rye is considered to be the species most Al tolerant in the Triticeae tribe (Miftahudin et al. 2002) and genes with a larger effect on Al tolerance were located on chromosomes 6RS (*Alt1*), 3RL (*Alt2*), and 4RL (*Alt3*) (Miftahudin et al. 2002; Gallego et al. 1998; Gallego and Benito 1997; Aniol and Gustafson 1984). Because of the synte observed in the chromosomes of group four among the Triticeae members, there is evidence that the *Alt3* gene on chromosome 4R in rye, Alt_{BH} on chromosome 4DL in wheat, and *Alp* on chromosome 4H in barley are orthologous loci.

According to Borgonovi et al. (1987), Al tolerance in sorghum is a trait controlled by few major genes, with emphasis of dominance effects. Recently, a gene with a major effect on Al tolerance, Alt_{SB}, was located in the terminal portion of sorghum chromosome 3 (Magalhães et al. 2007), accounting for up to 80 % of the phenotype variation in the trait. The large phenotypic diversity in Al tolerance observed among 13 sorghum lines, appears to be related to an allelic series in the Alt_{SB} locus as well as to the presence of other distinct genes (Caniato et al. 2007). The Alt_{SB} gene is a member of the MATE family (Multidrug and Toxin Compound Extrusion Family), which encodes a membrane transporter and mediates the exudation of citrate in the roots, induced in the presence of Al. This response was observed in a tolerant line, unlike the sensitive isogenic line, which presented a dramatic inhibition of root growth.

On the other hand, Al tolerance in maize has a complex inheritance, since progenies derived from crosses between tolerant and sensitive lines show continuous frequency distributions under Al stress (Magnavaca et al. 1987; Sawazaki and Furlani 1987). Magnavaca et al. (1987) reported a predominance of additive effects in the genetic variation associated with Al tolerance in maize. However, Moon et al. (1997) identified a gene with partial dominance (*ALM1*) responsible for tolerance to Al toxicity in this species, with the favorable allele identified in a line derived from a somaclonal variant of the "cateto" race (Cat-100-6). Subsequently, Sibov et al. (1999) mapped two QTL, called *ALM1* and *ALM2*, respectively, on chromosomes 6 and 10, which are involved in the genetic control of Al tolerance in maize. Ninamango-Cárdenas et al. (2003) mapped five Al tolerance QTL on maize chromosomes 2, 6, and 8, explaining about 60 % of the phenotypic variation of the trait. Of these QTL, only that on chromosome 6 was consistent between the two genetic mapping studies in maize, being equivalent to *ALM1*. Recently, Maron et al. (2009) characterized a member of the MATE family in maize, *ZmMATE1*, co-localized with the major Al tolerance QTL in the same region as chromosome 6. This candidate gene encodes a protein located in the plasma membrane that activates the citrate release in the root apex. *Arabidopsis* plants overexpressing this gene have Al tolerance increased. Krill et al. (2010) also identified candidate genes associated with Al tolerance in a maize association panel.

Similar to maize, Al tolerance in rice appears to be quantitatively inherited (Wu et al. 1997; Khatiwada et al. 1996). This observation was confirmed by mapping studies in crosses between different species of this genus (Nguyen et al. 2002; Wu et al. 2000).

The evaluation of diallel crosses in soybean revealed that additive effects are predominant in Al tolerance (Speher and Galwey. 1996; Speher 1995). Bianchi-Hall et al. (2000) found more than five QTL with minor effects, indicating that the control of Al tolerance in this species is quantitative.

Diallel crossing in alfalfa, a crop highly Al sensitive, also showed that non-additive effects were more important than the additive effects in the control of Al tolerance in this species (Campbell et al. 1993).

Fig. 8.2 Experiments conducted in nutrient solution for the evaluation of Al tolerance in sorghum. On the *left* is a detail of the tray containing inert material to maintain the seedlings suspended. In the *right* photograph, the length of the seminal root of the seedlings is measured. Photo provided by Lauro J. M. Guimarães

8.4 Stress Imposition and Evaluation of Al Tolerance

The adoption of reliable methods to evaluate aluminum tolerance is essential for success in the selection of superior genotypes. Plant responses to Al toxicity have been evaluated in controlled environments or under field conditions.

8.4.1 Evaluation in Nutrient Solution

The inhibition of root growth caused by Al is the basis of phenotypic analysis performed in nutrient solution. The main advantages of this method are the low cost, the high degree of experimental control, the high heritability, and the early evaluation of large numbers of genotypes. Furthermore, the method is not destructive and enables the transplant of superior genotypes to obtain seeds. Precise control of the stress level, the solution pH, and the availability of essential nutrients for seedling growth enable the isolation of Al toxicity from other factors that commonly occur in acid soils.

Protocols for conducting this kind of experiment in several species using different nutrient solutions are available. An example is the nutrient solution described by Magnavaca et al. (1987), which produces good results in maize and sorghum. In general, the seedlings are kept in trays with roots immersed in the nutrient solution, which must be constantly aerated (Fig. 8.2).

The intensity of stress is determined by the activity of free Al^{3+} in the nutrient solution and by the exposure time. These variables must be defined according to the species and the set of genotypes to be evaluated, since there are different levels of tolerance among and within species.

Different phenotype indexes based on the length of seminal roots can be used for selection, physiological, and molecular purposes. Two widely used indexes are

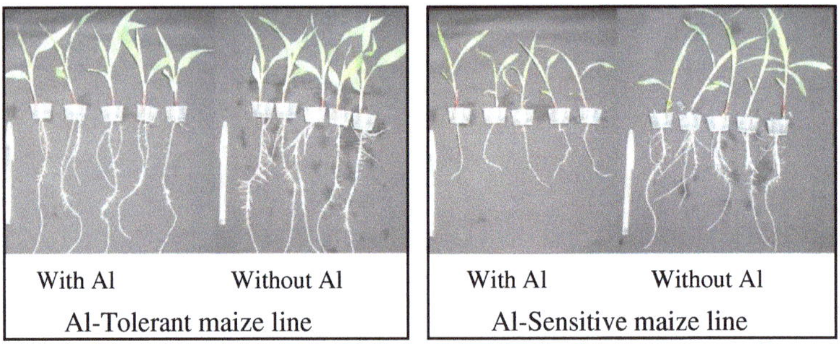

Fig. 8.3 Maize genotypes tolerant (*left*) and sensitive (*right*) to Al, evaluated in nutrient solutions with (39 μM of activity of Al) and without Al. Photo provided by Lauro J. M. Guimarães

based in the growth of the seminal root: the net seminal root length in the presence of aluminum (NSRL) and the relative seminal root length (RSRL). The NSRL is obtained by the difference between the final (FSRL) and initial (ISRL) seminal root lengths in the solution containing toxic level of Al, while the RSRL is the ratio of the net growth under conditions of Al stress to the net growth in the absence of Al.

The extent of root growth under the control condition (without Al) is essential to avoid interferences from different rates of root growth among genotypes. This control allows obtaining the RSRL, which is a phenotypic index of Al tolerance adjusted for the rate of root growth. Parentoni et al. (2003), using nine inbred maize lines and their 36 diallelic hybrids, concluded that the inclusion root growth data obtained without Al to generate the RSRL index, increased considerably the contribution of additive effects compared to the dominance effects. Additionally, the use of RSRL improved the prediction of general combining ability for Al tolerance of the lines evaluated in hybrid combinations. Then, it is an important phenotypic index that can be used for selection of superior lines.

Paterniani et al. (2002) performed experiments with solution containing Al and used the relative index of tolerance to aluminum (RIT_{Al}). This index enabled the comparison of a series of genotypes with two standard genotypes, one sensitive and one Al-tolerant. These two genotypes assume RIT_{Al} values equal to 1 and 5, respectively, according to the expression:

$$ITR_{Al} = \left[\frac{(CLRS_i - CLRS_S)}{(CLRS_T - CLRS_S)} x^4\right] + 1$$

Examples of maize genotypes tolerant and sensitive to Al, based on the seminal root growth observed after 5 days of treatment in nutrient solution containing Al, are shown in Fig. 8.3.

Since Al affects root growth, it is expected that the number of secondary roots is also lower. In this case, a more accurate and comprehensive analysis of the root system, can add important information to the study of genes and mechanisms of Al tolerance in plants. More detailed characteristics of roots can be obtained using equipments and softwares for acquisition and processing of images. The WinRhizo and other systems, such as the RootReader2D proposed by Famous et al. (2010), for example, are useful in studies of root traits. The measurement of the length and total volume of the root system, as well as the classification of the roots according to the diameter can be performed with such systems.

Using analysis of root images, Famoso et al. (2010) found differences in the total length of the rice roots grown for 3 days in nutrient solution without and with aluminum. Although the seminal roots present similar lengths, there were lower number and length of secondary roots in rice sensitive genotypes when compared to the tolerant ones. In that study, the total length of the roots grown in the solution containing toxic levels of Al was only 54 % of the total growth observed under control conditions (without Al).

Other traits, such as dry weight of roots and shoots, total weight, and plant height, can also be used to characterize genotypes for their tolerance to Al.

8.4.2 Evaluation in the Field

Despite the convenience and speed, results obtained in nutrient solution cannot be extrapolated directly to field conditions because other factors affect the yield of crops cultivated in acidic soils. Since the final objective of breeding programs is better yielding, it is important to evaluate cultivar under field conditions. Evaluation of crops in soils with high Al saturation allows the breeder to verify the effectiveness of selection for Al tolerance. These trials should represent the areas and seasons for which future cultivars will be recommended. However, for preliminary testing of the effect of Al tolerance on grain yield, it is important to reduce the influence of other limitation factors in the experimental area. It can be done by fertilization, irrigation, and correct field practices, leaving only differences in the saturation with Al in the stressed and non-stressed experimental essays.

Spatial heterogeneity in experimental areas is a common fact and is related to the processes of soil genesis and multiple agricultural activities in these areas. Thus, detailed knowledge of the experimental field and reports of agricultural uses are critical points for more correct phenotyping sites allocation, in order to define the most uniform areas to conduct the trials. Soil sampling in small grids with georeferenced points enables the construction of detailed maps of physical and chemical soil characteristics (Viana et al. 2009). These maps can be used in the application of lime and fertilizer at variable rates according to the heterogeneity of the area. In addition, these soil data can be used as environmental cofactors in statistical models. If possible, the water soil content should be monitored with sensors to enable proper control of irrigation when it is needed, as suggested by Albuquerque et al. (2005).

The experiments should be installed in areas with contrasting conditions, including the control condition without Al toxicity and areas that have between 40 and 80 % Al saturation, depending on the species to be evaluated. The target Al saturation level for test in field can be obtained by applying limestone according to the soil pH and the objectives of the study. The crop yield reduction should be between 40 and 60 % under stress condition, compared to the control (without Al toxicity), for effective identification of tolerant genotypes.

8.5 Strategies and Methods of Selection/Breeding

To begin a breeding program for Al tolerance, it is important to know the genetic control of this trait, which is already available for some crops. Such information enables the selection of the most appropriate strategies for development of superior genotypes.

A key step is the formation of base populations, from which new lines, hybrids, or varieties will be derived. The choice of the parents must be the start point on the selection of superior genotypes under Al stress conditions. Depending on the goal, the sources can range from elite materials to wild materials.

When the inheritance of a trait is predominantly controlled by non-additive genetic effects, i.e., there are dominance effects, the development of hybrids is the method that leads to larger gains. This strategy is particularly interesting for allogamous (out-crossing) species, allowing the exploitation of the hybrid vigor in specific crosses. Hybridization is also widely used in autogamous (self-crossing) crops, especially for those in which systems of cytoplasmic male sterility are well characterized, such as sorghum and rice. In addition, recurrent selection methods can be used for population breeding, by recombing superior families to increase the frequency of favorable alleles in the new selection cycles (Borém and Miranda 2009).

In autogamous (self-crossing) species, whose inbred lines are used as commercial cultivars, if the trait has a predominance of non-additive genetic effects, evaluations in Al stress condition should be performed in later generations in which there are high rates of homozygosity. Thus, the dominance effects are dissipated, enabling more appropriate comparisons between progenies. To this end, methods such as SSD (single seed descent) and pedigree are suitable. However, for traits that have predominant additive effects, the selection can be performed at early stages of homozygosity. In this case, the method of the population within families is also suitable.

8.6 Biotechnology: QTLs, Marker-Assisted Selection, Transgenics

Some species, like sorghum, present monogenic inheritance of Al tolerance in some crosses (Magalhães et al., 2007). Al tolerance in wheat is a trait presented in low frequency (Garvin and Carver 2003), which is probably a condition inherent to this trait in cultivated species. In these cases, there is a high probability that the superior alleles will not be present in the improved germplasm. Thus, targeted introgression of favorable alleles can be done using backcrossing assisted by molecular markers.

Molecular markers for assisted introgression were described in sorghum, by Caniato et al. (2007) and are available. Based on the Al tolerance gene (*Alt*$_{SB}$), cloned in sorghum (Magalhães et al. 2007), polymorphisms SNP-type (single nucleotide polymorphisms) were converted into Sequence Tagged Site (STS) markers, which are allele-specific markers or flanking this gene. Those markers together with microsatellites distributed in the sorghum genome, are being used for introgression of superior alleles of the *Alt*$_{SB}$ gene in elite lines from Brazil (Embrapa) and Niger (Oliveira et al. 2010).

Since Al tolerance is a quantitative trait in maize, a combination of strategies for QTL introgression, complemented by early phenotyping of lines, may be interesting for increasing the chances of success in generating tolerant materials. Even considering the existence of a QTL with a major effect (Ninamango-cárdenas et al. 2003; Sibov et al. 1999) co-localized with genes homologous to the *Alt*$_{SB}$ gene (Maron et al. 2009), other genomic regions should be monitored on the basis of early phenotypic selection or genome-wide selection (GWS). In the case of phenotypic selection in nutrient solution, Parentoni et al. (2003) have reported a high correlation between performance *per se* of maize inbred lines and its general combination ability (GCA) evaluated in diallel crosses.

In the case of introgression of genomic regions associated with the trait, marker-assisted backcrossing is an important breeding strategy. In maize, for example, three cycles of marker-assisted backcrossing were sufficient to recover approximately 99 % of the recurrent genome (Morris et al. 2003). However, for quantitative traits, one must consider that the phenotype variance is partitioned among multiple QTLs, which may be influenced by the environment. In this case, other strategies like assisted recurrent selection or genome-wide selection can produce more satisfactory results than introgression of individual QTLs.

Genetic transformation is another strategy that has been widely used for validation of Al tolerance genes. Transgenic barley plants that overexpress the *ALMT1* wheat gene showed a significant increase in the rate of malate exudation, which lead to increased tolerance to aluminum (Delhaize et al. 2004). Magalhães et al. (2007) also demonstrated that *Arabidopsis* plants transformed with the *Alt*$_{SB}$ genes showed higher Al tolerance and citrate exudation than non-transgenic plants. These results show that overexpression of these heterologous genes confers increased tolerance to Al and suggest an additional strategy for crops that have limited genetic variability for this trait. However, it is important to prove that the new alleles for Al tolerance in transgenic plants can increased grain yield or adaptability of cultivars to acid soils in the field conditions.

Significant advances in the knowledge of physiological and molecular basis for aluminum tolerance were obtained with the cloning of genes of major effects, such as *ALMT1* in wheat (Sasaki et al. 2004) and *Alt*$_{SB}$ in sorghum (Magalhães et al. 2007), that are involved in the Al exclusion mechanism. In addition, new genes and QTL have provided important support for a broader understanding of other mechanisms involved in Al tolerance in plants. However, the availability of cultivars with higher levels of Al tolerance would have a significant gain in time and efficiency with the broad integration of molecular and physiological knowledge into the breeding programs.

References

Adámoli J, Macêdo J, Azevedo LG de, Netto J (1985) Caracterização da região dos Cerrados. In: Goedert WJ (ed) Solos dos Cerrados: tecnologias e estratégias de manejo. São Paulo: Nobel; Planaltina, DF: EMBRAPA CPAC. Cap. 2:33–74

Albuquerque PEP, Durães FOM, Gomide RL, Andrade CLT (2005) Estabelecimento de sítios-específicos experimentais visando imposição e monitoramento de estresse hídrico para fenotipagem de cereais. Sete Lagoas: Embrapa Milho e Sorgo. (Embrapa Milho e Sorgo. Circular Técnica, 61) p 10

Aniol A, Gustafson JP (1984) Chromosome location of genes controlling aluminum tolerance in wheat, rye and triticale. Can J Genet Cytol 26:701–705

Bianchi-Hall CM, Carter TE, Bailey MA, Mian MAR, Rufty TW, Ashley DA et al (2000) Aluminum tolerance associated with quantitative trait loci derived from soybean PI 416937 in hydroponics. Crop Sci 40:538–545

Borém A, Miranda GV (2009) Melhoramento de plantas, 5th edn. Editora UFV, Viçosa, p 529

Borgonovi RA, Schaffert RE, Pitta GYE (1987) Breeding aluminum-tolerant sorghums. In: Gourley LM, Salinas JG (eds). Soghum for acid soils. Cali, Colombia: CIAT. pp 271–292. Proceedings of a workshoop on evaluating sorghum for tolerance to Al-toxic tropic soils in Latin America.

Buol SW (2009) Soils and agriculture in central-west and North Brazil. Scientia Agricola 66(5):697–707

Campbell A, Xia ZL, Jackson PR, Baligar VC (1993) Diallel analysis of tolerance to aluminium in alfalfa. Euphytica 3(72):157–162

Cançado GM, Parentoni SN, Borém A, Lopes MA (2002) Avaliação de nove linhagens de milho em cruzamentos dialélicos quanto à tolerância ao alumínio. Pesquisa Agropecuária Brasileira 37(4):471–478

Caniato FF, Guimaraes CT, Schaffert RE, Alves VMC, Kochian LV, Borém A, Klein PE, Magalhães JV (2007) Genetic diversity for aluminum tolerance in sorghum. Theor Appl Genet 114:863–876

Degenhardt J, Larsen PB, Howell SH, Kochian LV (1998) Aluminum resistance in the Arabidopsis mutant alr-104 is caused by an aluminum-induced increase in rhizosphere pH. Plant Physiol 117:19–27

Delhaize E, Craig S, Beaton CD, Bennet RJ, Jagadish VC, Randall PI (1993) Aluminum tolerance in wheat (*Triticum aestivum*). I. Uptake and distribution of aluminum in root apices. Plant Physiol 103:685–693

Delhaize E, Ryan PR, Hebb DM, Yamamoto Y, Sasaki T, Matsumoto H (2004) Engineering high-level aluminum tolerance in barley with the ALMT1 gene. Proc Nat Acad Sci USA 101:15249–15254

Famoso AN, Clark RT, Shaff JE, Craft E, McCouch SR, Kochian LV (2010) Development of a novel aluminum tolerance phenotyping platform used for comparisons of cereal aluminum tolerance and investigations into rice aluminum tolerance mechanisms. Plant Physiol 153:1678–1691

Ferreira RP, Moreira A, Rassini JB (2006) Toxidez de alumínio em culturas anuais. São Carlos: Embrapa Pecuária Sudeste. (Embrapa Pecuária Sudeste. Documentos 63). p 35

Foy CD (1988) Plant adaptation to acid aluminum-toxic soils. Commun Soil Sci Plant Anal 19:959–987

Furukawa J, Yamaji N, Wang H, Mitani N, Murata Y et al (2007) An aluminum-activated citrate transporter in barley. Plant Cell Physiol 48:1081–1091

Gallego FJ, Benito C (1997) Genetic control of aluminum tolerance in rye (*Secale cereale* L.). Theor Appl Genet 95:393–399

Gallego FJ, Calles B, Benito C (1998) Molecular markers linked to the aluminum tolerance gene *Alt1* in rye (*Secale cereale* L.). Theor Appl Genet 97:1104–1109

Garvin DF, Carver BF (2003) Role of the genotype in tolerance to acidity and aluminum toxicity. In: Rengel Z (ed) Handbook of Soil Acidity. Marcel Dekker, New York, pp 387–406

Huang CF, Yamaji N, Mitani N, Yano M, Nagamura Y et al (2009) A bacterial-type ABC transporter is involved in aluminum tolerance in rice. Plant Cell 21:655–667

Iuchi S, Koyama H, Iuchi A, Kobayashi Y, Kitabayashi S et al (2007) Zinc finger protein STOP1 is critical for proton tolerance in Arabidopsis and coregulates a key gene in aluminum tolerance. Proc Nat Acad Sci USA 104:9900–9905

Khatiwada SP, Senadhira D, Carpena AL, Zeigler RS, Fernandez PG (1996) Variability and genetics of tolerance for aluminum toxicity in rice (*Oryza sativa* L.). Theor Appl Genet 93:738–744

Kochian LV, Hoekenga OA, Piñeros MA (2004) How do crop plants tolerate acid soils? Mechanisms of aluminum tolerance and phosphorous efficiency. Annu Rev Plant Physiol Plant Mol Biol 55:459–493

Kochian LV (1995) Cellular mechanisms of aluminum toxicity and resistance in plants. Annu Rev Plant Physiol Plant Mol Biol 46:237–260

Krill AM, Kirst M, Kochian LV, Buckler ES, Hoekenga OA et al (2010) Association and linkage analysis of aluminum tolerance genes in maize. PLoS ONE 5:e9958

Lagos MB, Fernandes MIM, Camargo CEO, Federizzi LC, Carvalho FIT (1991) Genetics and monosomic analysis of aluminum tolerance in wheat (*Triticum aestivum* L.). Brazilian J Genet 14:1011–1020

Liu J, Magalhães JV, Shaff J, Kochian LV (2009) Aluminum-activated citrate and malate transporters from the MATE and ALMT families function independently to confer Arabidopsis aluminum tolerance. Plant J 57:389–399

Magalhães JV, Liu J, Guimarães CT, Lana UGP, Alves VMC, Wang Y, Schaffert RE, Hoekenga OA, Piñeros MA, Shaff J, Klein PE, Carneiro NP, Coelho CM, Trick HN, Kochian LV (2007) A gene in the multidrug and toxic compound extrusion (MATE) family confers aluminum tolerance in sorghum. Nat Genet 39:1156–1161

Magnavaca R, Gardner CO, Clark RB (1987) Inheritance of aluminum tolerance in maize. In: Gabelman HW, Loughman BC (eds) Genetic aspects of plant mineral nutrition. Martinus Nijhoff Publishers, Dordrecht, pp 201–212

Maron LG, Piñeros MA, Guimarães CT, Magalhães JV, Pleiman JK, Mao C, Shaff J, Belicuas SNJ, Kochian LV (2009) Two functionally distinct members of the MATE (multi-drug and toxic compound extrusion) family of transporterspotentially underlie two major aluminum tolerance QTLs in maize. Plant J 61:728–740

Martin RB (1992) Aluminum speciation in biology. In: Chadwick DJ, Whelan J (eds) Aluminum in biology and medicine. Wiley, New York, pp 5–25

Miftahudin R, Scoles GJ, Gustafson JP (2002) AFLP markers tightly linked to the aluminum-tolerance gene *Alt3* in rye (*Secale cereale* L.). Theor Appl Genet 104:626–631

Minella E, Sorrells ME (1992) Aluminum tolerance in barley: Genetic relationships among genotypes of diverse origin. Crop Sci 32:593–598

Minella E, Sorrells ME (1997) Inheritance and chromosome location of Alp, a gene controlling aluminum tolerance in 'Dayton' barley. Plant Breeding 116:465–569

Miyasaka S, Hawes MC (2001) Possible role of root border cells in detection and avoidance of aluminum toxicity. Plant Physiol 125:1978–1987

Moon DH, Ottoboni LMM, Souza AP, Sibov ST, Gaspar M, Arruda P (1997) Somaclonal-variation-induced aluminum-sensitive mutant from an aluminum-inbred maize tolerant line. Plant Cell Rep 16:686–691

Morris M, Dreher K, Ribaut JM, Khairallah M (2003) Money matters (II): Costs of maize inbred line conversion schemes at CIMMYT using conventional and marker-assisted selection. Mol Breeding 11:235–247

Nguyen VT, Nguyen BD, Sarkarung S, Martinez C, Paterson AH, Nguyen HT (2002) Mapping of genes controlling aluminum tolerance in rice: Comparison of different genetic backgrounds. Mol Gen Genet 267:772–780

Ninamango-Cárdenas FE, Guimarães CT, Martins PR, Parentoni SN, Carneiro NP, Paiva E (2003) Mapping QTLs for aluminum tolerance in maize. Euphytica 130:223–232

Ofei-Manu P, Wagatsuma T, Ishikawa S, Tawaraya K (2001) The plasma membrane strength of the root-tip cells and root phenolic compounds are correlated with Al tolerance in several common woody plants. Soil Sci Plant Nitrition 47:359–375

Oliveira BCFS, Moura PMA, Maciel BH, Guimarães CT, Caniato FF, Schaffert RE, Fonseca SC Jr, Jardim-Belicuas SN, Magalhães JV (2010) Utilização de marcadores moleculares na seleção assistida para tolerância ao alumínio em sorgo. Sete Lagoas: Embrapa Milho e Sorgo. (Embrapa Milho e Sorgo. Circular Técnica, 155). p 7

Parentoni SN, Alves VMC, Milach SK, Cançado GMA, Bahia Filho AFC (2001) Melhoramento para tolerância ao alumínio como fator de adaptação a solos ácidos. In: Nass LL et al (eds) Recursos genéticos e melhoramento. Rondonópolis, Fundação MT, pp 783–852

Parentoni SN, AlvesVMC, Oliveira AC, Magalhaes J, Coelho AM, Schaffert E, Guimaraes CT, Gama EEG, Godoy CL, Santos MX, Guimaraes PEO, Pacheco CAP, Meirelles WF, Ribeiro PH.E, Pitta GVE, Bahia Filho AFC (2003) Genetics of aluminum tolerance in maize evaluated in nutrient solution with and without control experiments. In: International symposium on plant breeding, Mexico, 2003. Book of abstracts. Mexico: CIMMYT p 60. Editado por Hallauer AR. e Magalhaes JV

Paterniani MEAG, Furlani PR (2002) Tolerância à toxicidade de alumínio de linhagens e híbridos de milho em solução nutritiva. Bragantia 61(1):11–16

Piñeros MA, Magalhaes JV, Alves VMC, Kochian LV (2002) The physiology and biophysics of an aluminum tolerance mechanism based on root citrate exudation. Plant Physiol 129:1194–1206

Samac DA, Tesfaye M (2003) Plant improvement for tolerance to aluminum in acid soils-A review. Plant Cell Tissue Org Cult 75:189–207

Sanchez PA, Salinas JG (1981) Low-input technology for managing oxisols and ultisols in tropical America. Adv Agron 34:279–406

Sasaki T, Yamamoto Y, Ezaki B, Katsuhara M, Ahn SJ, Ryan PR, Delhaize E, Matsumoto H (2004) A wheat gene encoding an aluminum-activated malate transporter. Plant J 37:645–653

Sawazaki E, Furlani PR (1987) Genética da tolerância ao alumínio em milho cateto. Bragantia 46:269–278

Sibov ST, Gaspar M, Silva MJ, Ottoboni LMM, Arruda P, Souza AP (1999) Two genes control aluminum tolerance in maize: Genetic and molecular mapping analysis. Genome 42:475–482

Speher CR (1995) Diallel analysis for mineral element absortion in trpical adapted soybeans (*Glycine max* (L.) Merril). Theor Appl Genet 90:707–7013

Speher CR, Galwey NW (1996) Diallel analysis for aluminum tolerance in trpical soybeans (*Glycine max* (L.) Merril). Theor Appl Genet 92:267–272

Stølen O, Andersen S (1978) Inheritance of tolerance to low soil pH in barley. Hereditas 88:101–105

Suhayda CG, Haug A (1985) Citrate chelation as a potential mechanism against aluminum toxicity in cells: The role of calmodulin. Can J Biochem Cell Biol 63:1167–1175

Tang Y, Sorrells ME, Kochian LY, Garvin DG (2000) Identification of RFLP markers linked to the barley aluminum tolerance gene *Alp*. Crop Sci 40:778–782

Taylor GJ (1988) The physiology of aluminum tolerance in higher plants. Commun Soil Sci Plant Anal 19:1179–1194

Viana JHM, Resende ÁV,Guimarães CT, Parentoni SN (2009) Avaliação da variabilidade espacial do solo em experimentos de eficiência nutricional em milho, conduzidos em área com baixos teores de nutrientes: um estudo de caso. Sete Lagoas: Embrapa Milho e Sorgo. Boletim de Pesquisa e Desenvolvimento, 18) p 19

Wu P, Liao CY, Hu B, Yi KK, Jin WZ, Ni JJ, He C (2000) QTLs and epistasis for aluminum tolerance in rice (*Oryza sativa* L.) at different seedling stages. Theor Appl Genet 100:1295–1303

Wu P, Zhao B, Yan J, Luo A, Wu Y, Senadihra D (1997) Genetic control of seedling tolerance to aluminum toxicity in rice. Euphytica 97:289–293

Xia JX, Yamaji N, Kasai T, Ma JAF (2010) Plasma membranelocalized transporter for aluminum in rice. Proc Nat Acad Sci USA 107:18381–18385

Chapter 9
Breeding for Heat-Stress Tolerance

Moacil Alves de Souza, Adérico Júnior Badaró Pimentel and Guilherme Ribeiro

Abstract Heat stress is one of the main abiotic stresses that affect the production of various crops in many parts of the world. High temperatures alter several metabolic processes reducing photosynthesis activity that results mainly in grain yield losses. In this context, identifying heat-stress tolerant genotypes with high yield potential could contribute to increasing food production. This chapter deals with aspects of breeding for heat-stress tolerance. First, heat stress and the tolerance mechanisms are characterized along with the main characteristics associated with heat stress, such as cell membrane thermal stability, canopy temperature depression, and some morphological characteristics. It then presents genetic control and selection environments for heat-stress tolerance, breeding methods, and selection strategies. The chapter ends with information on the use of biotechnological tools applied to breeding for heat-stress tolerance.

Keywords Crop breeding · Abiotic stress · Heat tolerance · Tolerance mechanism · Selection strategies

M. A. de Souza (✉) · A. J. B. Pimentel · G. Ribeiro
Federal University of Viçosa, Viçosa, Brazil
e-mail: moacil@ufv.br

A. J. B. Pimentel
e-mail: adericojr@yahoo.com.br

G. Ribeiro
e-mail: guilherme.tche@gmail.com

R. Fritsche-Neto and A. Borém (eds.), *Plant Breeding for Abiotic Stress Tolerance*,
DOI: 10.1007/978-3-642-30553-5_9,

9.1 Introduction

In spite of the exceptional advance in food production that contradicts the pessimistic predictions of shortage in the last century, there are still populations that live in hunger in some regions of the world, for example, countries in Africa and Asia.

At the beginning of the third millennium, the possibility of a world food crisis has arisen because of the growing consumption due to the exponential increase in the human population. Starting in 2005, world stocks of wheat and rice have been reduced to levels below safety levels that correspond to about 30 % of global consumption. This alarming situation has become a general concern of governmental and non-governmental authorities who seek alternatives to increase agricultural production before the imminent food shortage.

Balance between food production and consumption will be possible with improved productivity or expanded cultivated area. In the first case, gains are expected by genetic breeding and more efficient cropping techniques because it is still possible to exploit the existing genetic variability that has not yet been exhausted in most of the species of agronomic importance. Similarly, field management techniques have evolved greatly, but they could become even more efficient especially in aspects related to soil management, fertilization and weed, pest and disease control.

In spite of the competence of the scientists that act in the food production area, gain in productivity will not be sufficient to meet the food demand in the medium and long term. In this context, the expansion of several crops in non-traditional areas has become essential, but the environmental conditions of these areas, considered marginal, will present some stress factor that prevents full plant development.

Among the environmental factors, temperature is one of the main conditioners of plant growth and development. Heat conditions and water regimen are the climatic variables of greatest weight in agricultural zoning for the various cultivated species.

Deviation in temperature, below or above the optimum range, implies alterations in diverse metabolic processes, with qualitative and quantitative losses in the end product that in extreme cases can culminate in premature plant death.

A typical situation of damage caused by heat stress has been reported for the wheat crop in the central region of Brazil. In this region, there can be up to 50 % reduction in wheat grain yield between crops sown at the end of the summer and those sown in winter (Souza and Ramalho 2001).

In addition to the restrictions of the marginal areas to the cropping of many species, another complicating factor for food production is the global climatic change. Studies by scientists of the Intergovernmental Panel on Climate Change (IPCC) indicated the increase in global temperature as a threatening factor to cropping several agricultural species that could intensify food shortage in the most vulnerable parts of the planet (Assad et al. 2008). These scientists emphasized that increase in temperature, as a result of global warming, could cause excessive losses in grain harvests and profoundly alter the geography of agricultural production in Brazil.

In the face of the challenge of maintaining food production compatible with the growth in demand, genetic breeding for heat stress conditions is one of the most rational strategies. With the development of heat-stress tolerant cultivars, regions considered economically unsuitable will be incorporated into the productive system.

Given the need to overcome the limits imposed by heat stress, this chapter will present aspects regarding the genetic breeding of plants, focusing on the effects of temperature on some cultivated species and the different breeding strategies that are used for this purpose.

9.2 Heat-Stress Characterization

Heat stress is defined as increase in temperature above the critical value for a period of time sufficient to cause irreversible damage to plant growth and development. The base temperature or tolerance limit of the plant varies among species, among genotypes of the same species, and the phenotypic phases of the same genotypes.

A transitory 10–15 °C rise in temperature above the environmental temperature is considered heat shock or heat stress. However, heat stress is a complex function of the intensity, duration, temperature increase rate, and the sensitivity of each development phase of the plant.

High daily temperatures can have direct harmful effects associated to high tissue temperature or indirect effects associated to water shortage in the plant that occurs due to the high evapotranspiration rate (Hall 2011b). Its effect can be observed at night in the reduction of net photosynthesis caused by the high respiration rate and flower abortion and pollen sterility in some species.

During germination and emergence, high temperature can decrease or totally inhibit germination, reducing the stand and consequently the final crop yield. In other development phases, high-temperature damages the photosynthesis apparatus and affects respiration, water ratios, cell membrane stability, hormone levels, and the primary and secondary metabolites (Wahid et al. 2007).

Excessive heat, in addition to inducing quantitative and qualitative losses in crop production, shortens the duration of the cycle, reduces flower fertility percentage, accelerates the swelling period, decreases the mean grain weight, and increases fruit fall. In foods consumed fresh, for example fruits and vegetables, heat stress causes morphological defects that, in the eyes of the consumer, mean quality loss.

In wheat, high temperatures during grain swelling alter the protein composition and starch/protein ratio that interfere directly in the physical and chemical properties of the flour. Cowpea in high daytime temperatures can present asymmetrical cotyledons, pigmentation loss in the seed coat, and are rejected by consumers. The surface and internal tissues of tomato and citrus fruit can be damaged by a combination of high temperature and intense solar radiation (Hall 2011b). Benites (2007) reported a series of physiological disturbances in potato tubers caused by heat such as uneven growth, splits, internal cavities, alteration in the internal coloring, and necrosis.

Bearing in mind the diverse phases, means, and mechanisms by which excessive heat can affect cultivated species, Hall (1990) stated that the conception of a breeding program to develop heat-stress-tolerant cultivars should answer the following questions:

(i) What is the variation in the day and night temperature in the environments where the cultivars will be used and at what is the yield reduction intensity?
(ii) At which plant development phase is it more sensitive to high temperature?
(iii) How is heat-stress tolerance inherited? Is there another characteristic associated with heat-stress tolerance by genetic linking or pleitropy that influences the adaption or adjustment of the cultivar?

9.3 Heat-Stress Mechanisms and Associated Characteristics

In breeding programs, thousands of lines are assessed annually in order to find the one that presents the best genetic combination, generally determined by productivity. Thus, new cultivars are developed without understanding the impact of environment on the selection process of the diverse mechanisms related to the tolerance in the new cultivar.

9.3.1 Tolerance Mechanisms

Understanding and assessing the biochemical and physiologic mechanisms at the cell, molecular, and morphological levels, as complement to the traditional selection methods, are alternatives to increase progress in breeding. Using these tools prevents heat-stress tolerance genes from being lost during a selection process based only on yield.

The importance of selection based on several traits was emphasized by Giordano et al. (2005). These researchers assessed tomato lines for cropping in tropical and equatorial regions with high temperatures and reported that several traits associated to heat-stress tolerance are controlled by different genetic mechanisms. Thus, selection based on the greatest possible number of traits associated with heat-stress tolerance has been an efficient strategy.

Many physiological mechanisms can contribute to heat-stress tolerance in the field and according to Wahid et al. (2007), they can be classified as:

(i) *tolerance mechanisms*: higher photosynthesis rate, stay- green, cell membrane thermal stability, and heat shock proteins; and
(ii) *escape mechanisms:* canopy temperature depression (CTD) and earliness.

Although they are well established, the methodologies that involve physiological characteristics have resulted in little progress in detecting genotypic differences. The factors that hinder the use of these traits as selection criteria include variations due to the experimental protocols and the difficulty of association when interpreting the results as mere physiological processes or as processes that interfere in the final yield (Hall 1992).

It should be considered that physiological methods are not always viable when there are many populations or plants to be assessed, a fairly common situation in most breeding programs.

9.3.2 Characteristics Used in Breeding Programs

There are several methodological options to assess heat stress in natural or controlled environments. The most used characteristics will be described and associated with heat stress.

9.3.2.1 Cell Membrane Thermal Stability

Cell membrane thermal stability has been much used in heat stress studies. Its functionality in stress from high temperature is essential for the plants to adapt to these conditions. Membrane rupture enables the electrolytes to flee from the cells to the medium and their concentration can be quantified by electric conductance. Thus, greater electrical conductivity indicates greater heat-stress tolerance (Saadalla et al. 1990). Bearing in mind its inheritable nature and high genetic correlation with yield, Reynolds et al. (2001) stated that the potential was evident of applying cell membrane thermal stability to genetic breeding.

Reductions of up to 95 % were observed in rice grain yield as a result of high night temperatures (Mohammed and Tarpley 2009). According to the authors, this reduction is function of the electrolytic flight as a result of the decrease in membrane stability. Cytoplasm solute extrusion was attributed to the loss of membrane integrity. However, this phenomenon can also be caused by the inhibition of enzymes associated to the membranes, that maintain the chemical gradient in the cell (Mitra and Bhatia 2008). In this sense, in wheat, Blum et al. (2001), using a Recombinant Inbred Line (RIL) population, observed the association between the two characteristics. However, the coefficients of correlation obtained ranged from 0.5 to 0.6, indicating that membrane stability and grain production are not strongly associated. The authors concluded that cell membrane thermal stability per se could not be used as an exclusive selection criterion for heat stress in wheat genetic breeding.

9.3.2.2 Chlorophyll Fluorescence

The chlorophyll fluorescence technique has been adopted to quantify the effects of high temperatures on plants. According to Moffatt et al. (1990), this technique has been used to assess between and within tolerance in several species, based physiologically on the damage caused by high temperatures in the thylakoid tissues and related reactions with photosynthesis that alter the quantity of light absorbed, that is, transduced to photosystem II (FSII) for photosystem I (FSI). The quantity of chlorophyll fluorescence, that indicates the thylakoid membrane integrity and the relative efficiency of electron transport from FSII to FSI, is determined using a portable apparatus called a fluorometer, described by Moffatt et al. (1990).

Petkova et al. (2007) used chlorophyll fluorescence to assess common bean genotypes for heat-stress tolerance. Based on the damage caused to the photosynthesis apparatus and heat stress conditions in the field, the genotypes assessed could be grouped in as stress susceptible, intermediate, and tolerant. According to these authors, chlorophyll fluorescence can be used successfully under field conditions in breeding programs for heat tolerance.

9.3.2.3 Canopy Temperature Depression

The difference between the air and leaf temperature is defined as CTD. Infrared thermometry is one of the techniques used to measure the CTD of the plants and is much used for studies on drought tolerance. Its principal is the fact that plants with greater heat-stress tolerance can maintain their organ temperature, respiration, and transpiration activities at normal levels, even under stress conditions.

According to Reynolds et al. (2001), the use of CTD can be limited by the influence of environmental factors, because it involves several physiological processes, including water availability in the soil, air temperature, relative humidity, and radiation incidence. This characteristic is more suitable for superior line selection in warm environments with low relative air humidity, than where high evaporative demand leads to leaf cooling to below the environment temperature. This permits that the differences among the genotypes are detected fairly easily. However, in high humidity environments, these differences are not easily detected because the effect of evaporative leaf chilling is negligible.

Another concern when using CDT is that measuring the canopy temperature requires similar genotypes plots that is only possible in relatively advanced generations in the breeding program and it could not be used in early generations in individual plant selection.

9.3.2.4 Triphenyltetrazolium Test

Another physiological technique to quantify heat stress on cells is based on 2,3,5-triphenyltetrazolium chloride (TTC) reduction by dehydrogenase respiratory enzymes, that is, TTC assesses the mitochondria electron transport chain. The quantity of reduced TTC serves as indication of the level of mitochondrial respiration, determined by spectrophotometry and it reflects the relative viability of the cell (Porter et al. 1994).

In wheat there is a linear association between heat tolerance at the seedling stage and the flowering stage, with correlation values of 0.72 for TTC and 0.854 for cell membrane thermal stability (Fokar et al. 1998). Thus heat-stress tolerance can be predicted in adult plants by TTC or by cell membrane thermal stability, using seedlings in controlled environments saving time and space.

9.3.2.5 Morphological Characteristics Associated with Heat-Stress Tolerance

In addition to the biochemical and physiological characteristics, several morphological characteristics are used to measure the effect of heat on plants which are directly associated to the production components. Some of these potentially useful characteristics for heat-stress assessment include visual aspects, such as plant vigor, leaf senescence, soil covering, and stay-green. Others can be quantified or measured, such as development stage duration, number of emerged seedlings, tillering capacity, mean grain weight, number of grains per ear, harvest index, and yield.

Wheat cultivars with greater heat-stress tolerance have high swelling rates when submitted to heat-stress conditions. This suggests that the swelling rate and grain weight might be used as selection criteria for tolerant genotypes (Dias and Lidon 2009). Traits such as number of grains per ear, biomass, harvest index, and hectolictric weight can be used as potential selection criteria for grain production under high temperatures, due to the high correlation values observed (Yildirim and Bahar 2010).

In heat-sensitive species during reproductive development, but without photosynthesis restrictions, selection for flowering and fructification characteristics can be efficient in increasing heat-stress tolerance (Hall 1992). In cowpea breeding, high night temperatures reduced grain yield due to flower abortion and low pod setting indexes. Under this condition, selection based on flower number and quantity of well-formed pods under stress conditions has been efficient in developing tolerant cultivars.

In some cases, plant breeding can be used to neutralize the harmful effects of heat, induced by the acceleration of the reproductive development. In species with indeterminate growth habit, such as most legumes, cotton, and tomato, the length of the reproductive period can be altered by modifying the plant growth habit and production progression of plant nodes, branches, and reproductive nodes. In certain species, such as rice, sorghum, and wheat, plant and reproductive structure development is not very flexible and fast reproductive development due to high temperature substantially reduces the yield potential of these crops. As the reproductive period is short, the time available for photoassimilate accumulation and their translocation for grain development is also short (Hall 2011a). The translocation of photoassimilates that are fixed and stored before anthesis can be a mechanism of heat-stress tolerance in a determined crop submitted to heat stress during grain swelling.

Table 9.1 Expression of HSPs families in cereal plant organs

HSPs families	Organs/tissue	Cereal
HSP100	Leaf	Wheat
	Root	Wheat
	Seed	Wheat, corn and rice
HSP90	Seed	Corn
HSP70	Leaf	Wheat
	Root	Wheat and corn
	Seed	Wheat
HSP60	Leaf	Corn
	Root	Wheat, corn, barley and rye
	Seed	Corn
Small HSP	Leaf	Wheat, corn, barley and rye
	Root	Wheat and corn
	Seed	Wheat, sorghum and millet

Source adapted from Xu et al. (2011)

9.3.2.6 Heat Shock Proteins and Transcription Factors

Plants develop different adaptation mechanisms to heat stress, including alterations in protein metabolism. At high temperatures, many proteins are inhibited but the so-called heat shock proteins (HSPs) can have their synthesis increased. In addition to high-temperature the selective HSPs of some plants are induced in response to other stresses, such as water shortage, saline, and cold (Singh and Grover 2008). These proteins, together with their heat shock transcription factors (HSFs) are involved in molecular mechanisms for high temperature tolerance, acting as molecular chaperones to maintain protein partitioning homoeostasis. Due to these properties, the structural and metabolic integrity of the cells is maintained (Hong et al. 2009).

The HSPS protective function was proved by Burke (2001), who observed that *Arabidopsis thaliana* mutants that were unable to synthesize this class of protein or with cells where the synthesis of a specific HSP, HSP70, was blocked or inactivated, were more susceptible to lesions caused by heat.

The HSPs have been classified in five groups: HSP100, HSP90, HSP70, HSP60, and small HSP. Some HSFs have been identified and characterized associated to heat stress in cereals such as rice, wheat, corn, sorghum, rye, barley, and oats (Table 9.1), in addition to perennial species such as *Agrostis stolonifera, Festuca* sp., and *Dactylis glomerata*.

9.4 Genetic Variability and Control of Heat-Stress Tolerance

The existence of genetic variability in heat stress is an indispensable factor for the development of more tolerant cultivars. Similarly, knowledge of how the traits involved in heat-stress tolerance are inherited is very important for a successful breeding program. The best strategies are determined from these data to obtain, conduct and select the genetic material to be exploited.

Differences in heat-stress tolerance among wheat cultivars and segregate populations in the germination phase were reported by Cargnin et al. (2006). This statement enables selection under controlled conditions in the first seedling phases. For the potato, Benavente et al. (2011) reported that family selection in the seedling generation and first clonal generation for specific weight and in the first clonal generation for yield was efficient, under both cool and high temperatures, compared with selection in later generations. These results save time and resources in the breeding programs for greater heat-stress tolerance for this species.

Souza and Ramalho (2001) used hybrid wheat populations obtained in a partial diallel and observed that the genotypes assessed differed for the degree of heat-stress tolerance and that the additive and non-additive effects were important in the genetic control of grain production in the presence of heat stress. However, I-brahim and Quick (2001) observed that heat-stress tolerance, translated by membrane stability, was conditioned in great part by additive genetic effects and there is a significant maternal effect.

Blum et al. (2001) obtained relatively high heritability values when they assessed the heat-stress tolerance in wheat by grain yield and cell membrane thermal stability under heat-stress conditions, using a RIL population. These authors obtained $h^2 = 0.67$ for grain yield and $h^2 = 0.74$ for cell membrane thermal stability.

Using the estimates of additive genetic variants, variance due to the deviations of dominance and environmental factors, Fokar et al. (1998) reported that the greatest mean genetic variation by TTC in wheat can be explained by additive genetic variation and the heritability of this characteristic was 0.89.

9.5 Selection Environment for Heat-Stress Tolerance

One of the greatest difficulties in genetic breeding for heat-stress tolerance is environmental diversity because other factors interact with temperature and help to reduce the expected genetic gains.

Traditionally, the most used selection strategy has been assessing the genotypes in the environment with heat stress and selecting those that have yield greater than the best performing cultivars under these conditions. This supplies a direct measure of the heat-stress tolerance.

Direct selection under field conditions is usually difficult because some uncontrollable factors can affect trial precision and repeatability. These factors are:

(i) occurrence of other types of biotic and abiotic stress that affect assessment precision;
(ii) variation in stress intensity and duration in the cropping environments over the selection years;

(iii) variation in temperature in the same development phase between genotypes with different cycles; and
(iv) differences in sensitivity to heat stress at different development phases.

In spite of these problems, heat-stress tolerant genotypes can be successfully selected in the field, based on morphological traits. This strategy was used successfully in legumes, cotton, and tomato (Hall 2011a).

As an alternative to this breeding strategy in natural environments, selection can be made in controlled environments at the different plant development phases. These environments have the advantage of isolating the effect of heat stress from other external factors and maintaining constant stress intensity and duration over successive selection stages. However, it has the disadvantage of not being able to allow natural selection for other factors that interact with the heat-stress tolerance mechanisms under field conditions and is also more expensive.

Using the field or controlled environment depends on the development phase of the plant under assessment. If assessment and selection are made at the seedling, germination, or emergence phase, it is easier to maintain the soil temperature and assess the plant in a nursery. In some cases, the assessment and selection processes can be carried out in laboratories with controlled temperature during germination and seedling development.

For species with heat-sensitive root systems, it is advisable to use soil bed instead of pots for selection. The bed temperature is buffered, while pots are subject to overheating and consequently, the results obtained are not in line with the field conditions (Hall 1992).

The selection environment can differ regarding stress intensity under which selection is made. Some breeding programs prioritize selection under extreme temperature conditions to improve the distinction between genotypes with different degrees of heat-stress tolerance. However, this strategy can reduce the manifestation of genetic variability for exploitation. Furthermore, natural selection under these conditions can harm some relevant traits and favor undesirable characteristics in commercial production environments (Hall 1992).

Alternatively, genotypes can be assessed under cropping conditions with and without stress, and it is further possible, in the latter case, to assess in environments with graded stress levels.

In these situations, the genotype x environment interaction (GxE) is common, as reported by Oliveira (2008) in the wheat crop, where the Spearman coefficient of correlation for grain yield, among environments with and without stress, was low in the families ($r = 0.35$) and in the parents ($r = 0.24$). This indicated that the best families in the environment are not the best in another environment and that the genes are expressed differently in function of increase in temperature.

The breeder criterion, selecting the most heat-tolerant genotypes, can be obtained using the means of joint analysis or the means of the environment with stress. In any circumstance, the environment with heat stress cannot be excluded from the selection process otherwise the genetic gains for heat stress could be insignificant.

In the first case, genotypes can be identified with high stability in the face of environmental variations. This type of selection would be indicated if the greatest objective were not obtaining heat-stress tolerant plants for cropping in summer conditions but rather plants to be cropped under cool temperature conditions, but that can support higher temperatures if they occur.

Working with potato, Menezes et al. (2001) were successful with mean-based selection even in the presence of the GxE interaction, and identified superior parents and families in environments with and without stress, bearing in mind that these interactions represented the average performance of a large number of genotypes.

In the second case, the GxE interaction can be capitalized by selecting genotypes highly adapted to the stress condition. However, genotypes that are highly adapted to specific environments (with stress) but do not respond to environment improvement (without stress) are not desirable. This type of selection would be recommended only in cropping situations where stress occurrence is a constant and not a possibility.

The selection index is an alternative for selecting genotypes assessed in various environments, bearing in mind that it exploits the combination of multiple data contained in the experimental unit that allows selection based on more than one environment of interest.

Cargnin et al. (2007) used simultaneous selection in environments with and without stress to optimize gains under these two conditions and reported that the highest gains estimate were obtained for direct selection in each cropping environment, always with negative gain estimates for the indirect response in another environment. With simultaneous selection for environments, all the indices were more suitable for selection, because they registered higher total gains that were better distributed between the two environments. The index based on "desired gains" permitted the highest gains in the three assessment situations, that is, grain yield considered as principal in the two environments, in the summer and winter environments (Table 9.2).

An example of the diversity of environments that present heat stress and the influence of other climatic variables can be observed in the CIMMYT (International Maize and Wheat Improvement Center) breeding program for heat-stress tolerance in wheat in cooperation with researchers from other countries. In these studies, carried out in more than 40 locations, in various years, it was verified that most of the interaction between high temperature locations and genotypes was influenced by the relative air humidity. Based on these results, the experimental locations were stratified into groups, with specific breeding strategies (Reynolds et al. 2001).

9.6 Breeding Methods and Selection Strategies

When adopting any breeding method for a determined crop, the type of reproduction of the species should be considered, that is, whether it is self-pollinating, cross-pollinating, or asexual. For each crop, the objective can be the development

Table 9.2 Selection gain (SG) in wheat grain yield with direct and indirect selection (SDI), by the Smith (1936) and Hazel (1943) (SH), Williams (1962) (BW) and Pesek and Baker (1969) (PB) indexes, with economic weight and desired gain (EW/DG)

Índex	EW/DG	Situation	SG %		Total
			Summer	Winter	
SH	1	I[a]	9.4	34.1	43.5
		II[b]	30.2	−0.5	29.6
		III[c]	−5.2	37.7	32.6
SH	CVg[d]	I	7.6	34.9	42.6
		II	30.2	−0.5	29.6
		III	−5.2	37.7	32.6
BW	1	I	10.6	33.5	44.1
		II	30.2	−0.5	29.6
		III	−4.5	37.7	33.2
BW	CVg[d]	I	7.6	34.9	42.6
		II	30.2	−0.5	29.6
		III	−4.5	37.7	33.2
PB	SDg[e]	I	20.0	25.5	45.5
		II	30.1	0.8	30.9
		III	−1.8	37.5	35.8
SDI		II	27.5	−2.8	24.7
SDI		III	−2.2	36.4	34.2

[a] Grain yield considered as principal in the two environments
[b] Grain yield considered as principal in the summer environment
[c] Grain yield considered as principal in the winter environment
[d] Coefficient of genotypes variation
[e] Genotypes standard deviation

of pure lines, hybrid cultivars, cross-pollination cultivars, or clones. Defining the cultivar type to be obtained, associated to the genetic control of the traits that are being bred, will determine the best strategy to obtain and select the bred material.

9.6.1 Germplasm Introduction

Movement of seeds or plant parts among countries and regions has been a very common practice since early times. Germplasm introduction, defined here as lines and cultivars for direct use or populations for later selection, is considered as a breeding method because of the effective contribution of this practice to improving production in a determined region.

In non-traditional cropping areas, such as those limited by adverse climatic conditions, introducing cultivars or lines is an important alternative to expand the new crop, because it is a fast method of recommending a cultivar since superior genotypes can be made available immediately, similar to those that would be developed by an *in loco* breeding program. The success of using of this method can be optimized when the introduced germplasm comes from regions with edaphoclimatic conditions similar to those where the crop is being introduced. An example of successful use of this strategy was reported by Giordano et al. (2005), who assessed tomato lines in the region of Brasília derived from two introduced populations from Roraima, North Brazil. Heat-stress-tolerant lines were identified in this study with good yield and fruit quality.

Wheat line introduction has been used in the central region of Brazil, considered as a marginal area for the cereal, where high temperatures are the main limitation to crop development. In this region, that includes the states of Minas Gerais, Goiás, Mato Grosso, Bahia, and the Distrito Federal and in the states of Paraná, São Paulo, and Mato Grosso do Sul, 33 % of the 36 cultivars recommended from 1922 to 1997 were introductions from Mexico, 42 % from other Brazilian states, and only 25 % were developed by the research institutions of the region (Sousa 1997).

9.6.2 Breeding by Hybridization

When two or more desirable phenotypes are in different cultivars, an alternative is to combine them in a new line by hybridization. If the trait to be improved is quantitatively inherited, for example, heat-stress tolerance, the favorable alleles are certainly dispersed in different lines or cultivars; in this situation planned hybridization, to involve the different genetic sources, is essential to obtain expressive gains.

9.6.2.1 Parent Selection and Obtaining the Segregant Population

For traits controlled by few genes and little influenced by the environment, the choice of the parents to be crossed is fairly easy, it is enough for one of the two parents to be well adapted and the other to carry the alleles of interest. However, when traits with more complex genetic control are involved, parent selection is more difficult, because knowledge of their genetic complementariness or diversity is essential for successful breeding.

Using the mean as parent selection criteria is without doubt the most used method. For any quantitative trait, the parents should be chosen considering those that present superior mean for the desired trait. At this stage the heat stress assessment techniques, using physiological traits, can contribute greatly to parent selection.

Selecting parents based only on the mean does not ensure obtaining populations with sufficient variability for successful selection. In addition to being productive and adapted, the parents to be crossed should be genetically complimentary.

The parents chosen can be used in single crosses, although in some situations more complex crosses are needed, involving three or more parents. Complex or multiple crosses are an alternative that accelerate the combination process of

Table 9.3 Genetic gain prediction based on direct selection on wheat grain yield, the classic Smith and Hazel index, rank sum index, and Z standardized index

Selection strategy	Genetic gain (%)				
	GY[a]	EN[b]	GWE[c]	WMG[d]	HI[e]
Direct selection	18.8	6.1	7.1	2.2	8.46
Smith and Hazel index	18.8	5.6	8.4	3.5	10.72
Rank sum index	16.2	4.2	10.5	6.1	10.72
Z standardized index	16.9	5.6	10.1	5.0	10.72

Source adapted from Assis (2011)
[a] Grain yield
[b] Ear number
[c] Grain weigth per ear
[d] Weigth of 1000 grains
[e] Harvest index

several parents, because it takes a long time with bi-parental crosses to obtain a line and carry out later recombination.

In multiple crosses, the probability is very small of obtaining superior genotypes that reunite all the favorable alleles. In this circumstance a segregant population is required with a very large number of individuals that becomes unfeasible in practice. The alternative is to adopt recurrent selection to gradually accumulate, by recombination cycles, the desirable, and available alleles in different parents (Ramalho 1997).

For breeding for heat-stress tolerance, productivity in most cases is the main trait used for selection. Using production components that are associated with yield and higher heritability values can result in greater gains.

Selection indexes are an alternative in addition to direct selection and correlated response. They are defined as an additional trait, established by the optimum combination of several traits that allows effective simultaneous selection of multiple traits. With the use of the selection indexes, breeding is possible for a set of characteristics and according to Garcia and Souza Júnior (1999), they are suitable for recurrent selection programs.

In the case of selecting productive wheat progenies with greater heat-stress tolerance, gain prediction with truncated direct selection in grain yield was more efficient compared to selection indexes as shown by the results obtained by Assis (2011) (Table 9.3). However, the results obtained with the indices were also satisfactory.

9.6.3 Populational Breeding: Recurrent Selection

One of the alternatives that can be used to increase gains by selection is synthesizing populations with a wider genetic base and conducting them by recurrent selection.

Table 9.4 Genetic gain (GG) in grain yield in two recurrent selection cycles for heat-stress tolerance in wheat

Cultivation condition	GG (g/1,67 m^{-2})[a]		GG (%)[a]	
	Cycle I	Cycle II	Cycle I	Cycle II
With stress	93.66	98.69	39.42	34.37
Without stress	114.08	116.44	20.60	24.62
Average of environments	65.69	88.01	16.61	23.16

Source adapted from Machado et al. (2010)

[a] Estimated gain as deviation of the means of the 48 best families (20 %) compared to the mean of the common controls

Thus, plant breeders have emphasized populational breeding and quantitative traits, considering them directly responsible for increase in yield. In this sense, recurrent selection seems to be one of the most efficient methods in plant breeding, especially when quantitative hereditary characteristics are involved, for example abiotic stresses.

The following advantages of recurrent selection were highlighted by Geraldi (1997):

(a) greater genetic variability obtained by intercrossing of multiple parents;
(b) greater opportunity for recombinations because of successive crossings;
(c) greater efficiency in increasing the favorable gene frequency because the process is repetitive and accumulative;
(d) greater facility to incorporate exotic germplasm in the population.

Recurrent selection is characterized by the recombination of the best genotypes to form a bred population that maintains high genetic variability. It is a cyclic selection process of superior individuals or families within a population, followed by recombination to form a new population.

Recurrent selection has been widely used in corn breeding, a self-pollinating species. Its use has been widened successfully in recent decades in many countries in various self-pollinating species. In wheat, recurrent selection has been used with satisfactory results, as reported by Maich et al. (2000), who observed 15 % genetic progress in grain yield after two selection cycles. Genetic progress of 8.4 % was obtained in the third recurrent selection cycle for heat-stress tolerance in wheat, in research developed by the Wheat Breeding Program at the Federal University of Viçosa, in 2011. The efficaciousness of using this breeding strategy was evident from the gains observed under specific cropping conditions and in the mean of environments (Table 9.4).

The potato breeding program at the Federal University of Lavras has also been successful in developing heat-stress tolerance genotypes using recurrent selection. Benites and Pinto (2011) obtained expressive gains in tuber production with improved quality (Table 9.5).

Table 9.5 Genetic gain (GG) obtained in some potato traits under heat stress in Lavras

Generation	Tuber yield (g/plant)	Mean weight of large tubers (g)	% Large tubers	Tuber specific gravity	Tuber appearance
Cycle 1	37.19	13.38	40.80	0.48	17.19
Cycle 2	−5.76	1.36	0.66	0.40	3.07
Cycle 3	7.2	−1.62	−5.24	−0.05	−3.27
Total	37.85	12.99	32.40	0.83	16.64

Source adapted from Benites and Pinto (2011)

9.7 Biotechnology and Heat-Stress Tolerance

In classic breeding for heat-stress tolerant cultivars it is difficult to identify and use tolerant sources because of the limited genetic base of many cultivated species. Yield potential is hard to separate from heat-stress tolerance when assessing wild accessions and there is association among heat-stress tolerant genes and undesirable agronomic characteristics. Greater success can be obtained by integrating classic breeding and molecular biology. Using biotechnological tools allows the identification of candidate genes that once their function and association to determine molecular marker are proven, can be easily incorporated by assisted selection. Furthermore, introgression of genes that are known to be involved in the stress response, via genetic engineering, can be a faster breeding strategy to develop tolerant cultivars. However, this would be the only option when the genes of interest were derived from other sexually incompatible species, such as distant relatives, or from non-plant sources.

According to Wahid et al. (2007) in the last 20 years molecular marker-assisted selection and genetic transformation have contributed to understanding the genetic and biochemical base of plant tolerance to stress and in some cases has led to the development of tolerant plants. In this period, much research was carried out on different species to identify markers associated with different abiotic stresses, such as drought, salinity, and cold, while comparatively few studies were carried out to identify genetic markers associated to heat-stress tolerance. According to the same authors, the critical factor that limits using biotechnological tools, such as gene transformation and genetic expression techniques, in the case of heat stress, is that the factors that confer greater temperature tolerance in superior plants are not yet fully understood.

A way of identifying genes involved in heat stress is to submit the genotypes to adverse conditions, and identify, with the help of biotechnology, the genes that are expressed and thus select a group of candidate genes based on their probable function. In this context, QTL mapping can be used to identify candidate genes.

Mohammadi et al. (2008) detected three QTLs for heat-stress tolerance in wheat, measured by the stress susceptibility index (SSI) in 144 RILs. One was located on the 1B chromosome, strongly linked to the gwm190 marker and explained 44.3 % of the phenotypic variance. The second was located on chromosome 5B, close to the gwm133A and explained 27.3 % of the phenotypic variance. The third QTL was located on chromosome 7B, strongly linked to the gwm63B marker. These three QTL together explained 16.7 % of the phenotypic variance of the SSI, a very low value compared to that of each QTL individually, suggesting their statistical correlation. Much of the phenotypic variance explained and small confidence intervals indicate that the precision in locating the QTL detected is sufficient and that the linking between markers and QTL can be used in breeding programs.

Heredity and the minimum number of genes that confer heat-stress tolerance in common wheat were studied by Yang et al. (2002). Two markers, Xgwm11 and Xgwm293, were identified and were significantly associated with heat-stress tolerance. The QTL linked to the marker presented only additive genetic action and contributed 11 % to the phenotypic variation, while the QTL linked to the marker presented both additive and dominant genetic action and contributed 12 % to the total variation. These results showed that heat-stress tolerance in wheat is controlled by multiple genes and that microsatellite marker assisted selection might be useful in the development of bred cultivars.

Merlino et al. (2009) identified and mapped genes that codified HSPs in hexaploid wheat. According to the authors, these genes are distributed on the wheat genome as follows: Markers agl 136 and agl 147, similar to HSP70 found in barley, are located on the short arm of chromosome 6B. Marker agl 192, also identified as HSP70, was mapped on the region of the centromere of chromosome 1B. Markers agl 1977, agl 2012, agl 1984, and agl 1994, identified as HSP16.9 kDa, together with agl 2049, agl 2050, and agl 2066, identified as HSP17.5 kDa or HSP17.8 kDa, were all mapped on the chromosomes of group 3. The marker agl 1981, identified as rice HSP18, was mapped on the short arm of chromosome 4D.

In a study by Katiyar-Agarwal et al. (2003), the heat shock protein 101 (HSP101), responsible for transmitting heat tolerance in cells was successfully by Agrobacterium mediated transformation from *A. thaliana* to a rice cultivar. Good recovery was observed in the transgenic rice lines after being submitted to heat stress.

According to Singh and Grover (2008), the vast data on different HSP families provided by the availability of complete sequencing data of the *Oryza sativa* and *A. thaliana* genomes has facilitated identification and cloning of a large number of HSP genes in various species belonging to different taxonomic classes.

The availability of high density genetic maps and physical maps, EST's, genomic sequences, and mutants makes rice an excellent model to study heat-stress tolerance in cereals. The high level of syntheny and homology in the Poaceas family facilitates the transfer of QTL and identified candidate genes from rice to other cereals (Maestri et al. 2002).

Successful use of these biotechnological tools, however, does not only depend on identified genes that are related to heat stress. In addition to identifying and transferring these genes, assessing plants under stress conditions and understanding the physiological effects of the genes inserted at whole plant level are still great challenges to be overcome.

Thus, the use of biotechnological tools applied to breeding for heat-stress tolerance is only beginning and molecular marker technology and genetic transformation should contribute significantly to the development of plant tolerance to high temperatures in the future.

9.8 Conclusions

This chapter presented an approach to the development of heat-stress-tolerant genotypes, emphasizing the characterization of heat stress, identifying the most critical stages, and the main alterations caused in plants. It demonstrated that knowledge of characteristics used in the breeding programs has been shown to be an efficient indirect selection criterion in the development of heat-stress-tolerant genotypes.

Selection strategies with germplasm introduction and crossing of heat-stress-tolerant parents, to identify superior genotypes, together with the identification of selection environments that maximize the genetic gains are used successfully in genetic breeding for heat-stress tolerance further. In addition to the classic breeding techniques, studies at molecular level are useful to identify genes involved in heat stress and contribute to the development of tolerant plants.

References

Assad ED, Pinto HS, Zullo J Jr, Marin FR, Pellegrino GQ, Evangelista SR, Otavian AF (2008) Aquecimento global e anova geografia da produçao agrícola no Brasil. 1st edn. Brasília, Embaixada Britânica vol 1. p 82

Assis JC (2011) Progresso genético em três ciclos de seleção recorrente para tolerância ao calor em trigo. Universidade Federal de Viçosa, Viçosa, 74 pp (Tese de doutorado)

Benavente CAT, Pinto CAB, Figueiredo ICR, Ribeiro GHMR (2011) Repeatability of family means in early generations of potato under heat stress. Crop Breed Appl Biotechnol 11:330–337

Benites FRG (2007) Seleção recorrente em batata visando tolerância ao calor. Universidade Federal de Lavras, Lavras, p 90 (Tese Doutorado)

Benites FRG, Pinto CABP (2011) Genetic gains for heat tolerance in potato in three cycles of recurrent selection. Crop Breed Appl Biotechnol 11:133–140

Blum A, Klueva N, Nguyen HT (2001) Wheat cellular thermotolerance is related to yield under heat stress. Euphytica 117:117–123

Burke JJ (2001) Identification of genetic diversity and mutations in higher plant acquired thermotolerance. Physiol Plant 112:167–170

Cargnin A, Souza MA, Dias DCF, Machado JC, Machado CG, Sofiatti V (2006) Tolerância ao estresse de calor em genótipos de trigo na fase de germinação. Bragantia 65:245–251

Cargnin A, Souza MA, Machado CG, Pimentel AJB (2007) Genetic gain prediction for wheat with different selection criteria. Crop Breed Appl Biotechnol 7:334–339

Dias AS, Lidon FC (2009) Evaluation of grain filling rate and duration in bread and durum wheat, under heat stress after anthesis. J Agron Crop Sci 195:137–147

Fokar M, Nguyen HT, Blum A (1998) Heat tolerance in spring wheat. I. Estimating cellular thermotolerance and its heritability. Euphytica 104:1–8

Garcia AAF, Souza Júnior CL (1999) Comparação de índices de seleção não paramétricos para a seleção de cultivares. Bragantia 58:253–267

Geraldi IO (1997) Seleción recurrente em el mejoramiento de plantas. In: Guimarães EP (ed) Seleción recurrente em arroz. Centro Internacional de Agricultura Tropical, Cali, pp 3–11

Giordano LB, Boiteux LS, Silva JBC, Carrijo OA (2005) Seleção de linhagens com tolerância ao calor em germoplasma de tomateiro coletado na região Norte do Brasil. Horticultura Brasileira 23:105–107

Hall AE (1990) Breeding for heat tolerance—an approach based on whole-plant physiology. HortScience 25:17–19

Hall AE (1992) Breeding for heat tolerance. Plant Breed Rev 10:129–168

Hall AE (2011a) Heat stress and its impact. Plantstress. http://www.plantstress.com/Articles/index.asp. Accessed 26 Febr 2011

Hall AE (2011b) The mitigation of heat stress. Plantstress. http://www.plantstress.com/Articles/index.asp. Accessed 26 Febr 2011

Hazel LN (1943) The genetic basis for constructing selection indexes. Genetics 28:476–490

Hong B, Ma C, Yang Y, Wang T, Yamaguchi-Shinozaki K, Gao J (2009) Over-expression of AtDREB1A in chrysanthemum enhances tolerance to heat stress. Plant Mol Biol 70:231–240

Ibrahim AMH, Quick JS (2001) Genetic control of high temperature tolerance in wheat as measured by membrane thermal stability. Crop Sci 41:1405–1407

Katiyar-Agarwal S, Agarwal M, Grover A (2003) Heat-tolerant basmati rice engineered by over-expression of *hsp101*. Plant Mol Biol 51:677–686

Machado JC, Souza MA, Oliveira DM, Cargnin A, Pimentel AJB, Assis JC (2010) Recurrent selection as breeding strategy for heat tolerance in wheat. Crop Breed Appl Biotechnol 10:9–15

Maestri E, Klueva N, Perrotta C, Gulli M, Nguyen HT, Marmiroli N (2002) Molecular genetics of heat tolerance and heat shock proteins in cereals. Plant Mol Biol 48:667–681

Maich RH, Gaido ZA, Manera GA, Dubois ME (2000) Two cycles of recurrent selection for grain yield in bread wheat: direct effect and correlated responses. Agriscientia 17:35–39

Menezes CB, Pinto CABP, Lambert ES (2001) Combining ability genotypes for cool and warm seasons in Brazil. Crop Breed Appl Biotechnol 1:145–157

Merlino M, Leroy P, Chambon C, Branlard G (2009) Mapping and proteomic analysis of albumin and globulin proteins in hexaploid wheat kernels (*Triticum aestivum* L.). Theor Appl Genet 118:1321–1337

Mitra R, Bhatia CR (2008) Bioenergetic cost of heat tolerance in wheat crop. Curr Sci 94: 1049–1053

Moffatt JM, Sears RG, Paulsen GM (1990) Wheat high temperature tolerance during reproductive growth. I. Evaluation by chlorophyll fluorescence. Crop Sci 30:881–885

Mohammed AR, Tarpley L (2009) Impact of high nighttime temperature on respiration, membrane stability, antioxidant capacity, and yield of rice plants. Crop Sci 49:313–322

Mohammadi V, Zali AA, Bihamta MR (2008) Mapping QTLS for heat tolerance in wheat. J Agric Sci Technol 10:261–267

Oliveira DM (2008) Seleção em populações de trigo visando tolerância ao estresse de calor. Universidade Federal de Viçosa, Viçosa, 60 pp (Dissertação de Mestrado)

Pesek J, Baker RJ (1969) Desired improvement in relation to selection indices. Can J Plant Sci 1:215–274

Petkova V, Denev I, Cholakov D, Porjazov I (2007) Field screening for heat tolerant common bean cultivars (*Phaseolus vulgaris* L.) by measuring of chlorophyll fluorescence induction parameters. Hortic Sci 111:101–106

Porter DR, Nguyen HT, Burker JJ (1994) Quantifying acquired thermal tolerance in winter wheat. Crop Sci 34:1686–1689

Ramalho MAP (1997)Melhoramento do feijoeiro.In: Simpósio sobreatualizaçao em genética e684 melhoramento deplantas, Anais. Lavras, UFLA, pp 167–196

Reynolds MP, Nagarajan S, Razzaque MA, Ageeb OAA (2001) Heat tolerance. In: Reynolds MP, Ortiz-Monasterio JI, Mcnab A (eds) Application of physiology in wheat breeding. CIMMYT, México, pp 124–135

Saadalla MM, Quick JS, Shanahan JF (1990) Heat tolerance in winter wheat. II. Membrane thermostability and field performance. Crop Sci 30:1248–1251

Singh A, Grover A (2008) Genetic engineering for heat tolerance in plants. Physiol Mol Biol Plants 14:155–166

Smith HF (1936) A discriminant function for plant selection. Ann Eugen 7:240–250

Sousa CNA (1997) Relação das cultivares comerciais de trigo no Brasil de 1922 a 1997. EMBRAPA/CNPT, Passo Fundo p 46

Souza MA, Ramalho MAP (2001) Controle genético e tolerância ao estresse de calor em populações híbridas e em cultivares de trigo. Pesquisa Agropecuária Brasileira 36:1245–1253

Wahid A, Gelani S, Ashraf M, Foolad MR (2007) Heat tolerance in plants: an overview. Environ Exp Bot 61:199–223

Williams JS (1962) The evaluation of a selection index. Biometrics 18:375–393

Xu Y, Zhan C, Huang B (2011) Heat shock proteins in association with heat tolerance in grasses. Int J Proteomics 2011:1–11

Yang J, Sears RG, Gill BS, Paulsen GM (2002) Growth and senescence characteristics associated with tolerance of wheat-alien amphiploids to high temperature under controlled conditions. Euphytica 126:185–193

Yildirim M, Bahar B (2010) Responses of some wheat genotypes and their F_2 progenies to salinity and heat stress. Sci Res Essays 5:1734–1741

Chapter 10
Breeding Perennial Species for Abiotic Stress

Rinaldo Cesar de Paula, Nadia Figueiredo de Paula and Celso Luis Marino

Abstract In agriculture, the term "stress" is understood to mean any factor, whether biotic or abiotic, that is detrimental to the growth and productivity of plants. Managing stress caused by climatic events, especially in large perennial species occupying extensive cropping areas, is in most cases neither technically nor economically viable. In this chapter we will approach the breeding of perennial species tolerant to abiotic stress based on studies of *Eucalyptus* under water deficiency conditions, the most common stress factor in areas used for cultivating this genus in Brazil. The forestry breeding programs conducted in Brazil have been primarily directed toward growth traits, and more recently, wood quality. Indirectly, there has been selection, albeit unintentional, for adaptation to the environmental conditions under which commercial companies operate, and it can be assumed that drought-tolerance characteristics have not been fully explored in these programs. As new forest frontiers have opened up in regions with limited water resources, the search for productive and therefore drought-tolerant genotypes should be a key feature in breeding programs. Since the species concerned are perennials with a long life cycle, during their development the plants pass through innumerable situations that are unfavorable to growth. In contrast to short-cycle species, trees cannot avoid these conditions. This means that the mechanisms for withstanding abiotic stress factors in perennial species may be more complex and primarily aimed at species survival, to

R. C. de Paula (✉)
Universidade Estadual Paulista, Jaboticabal, Brazil
e-mail: rcpaula@fcav.unesp.br

N. F. de Paula
Faculdade de Tecnologia de Jaboticabal, Jaboticabal, Brazil
e-mail: nadiafp@hotmail.com

C. L. Marino
Universidade Estadual Paulista, Botucatu, Brazil
e-mail: clmarino@ibb.unesp.br

R. Fritsche-Neto and A. Borém (eds.), *Plant Breeding for Abiotic Stress Tolerance*,
DOI: 10.1007/978-3-642-30553-5_10, © Springer-Verlag Berlin Heidelberg 2012

the detriment of productivity. Understanding these mechanisms is essential for success of a forest breeding program for regions subject to drought.

Keywords *Eucalyptus* · Drought tolerance · Tree breeding · Water stress

10.1 Introduction

In agriculture, the term "stress" is understood to mean any factor, whether biotic or abiotic, that is detrimental to the growth and productivity of plants. The main abiotic stress factors on plants are those associated with soil conditions (aluminum toxicity, low pH, low fertility, etc.) and climate (unfavorable temperatures, radiation, relative humidity, and rainfall conditions). Edaphic stress factors can be manipulated and managed, within certain limits, in order to offset or minimize their effects using soil preparation techniques and fertilizers. This applies to both annual and perennial crops. However, managing stress caused by climatic events, especially in large perennial species occupying extensive cropping areas, is in most cases neither technically nor economically viable. Another important aspect to be taken into consideration is that the intensity and duration of stress factors are not predictable and can potentially cause enormous damage. Furthermore, the stress caused by abiotic factors can take the form of both excess and scarcity of a given resource, although scarcity is more common. In this chapter, the term "stress" is used as synonymous with "deficiency" or "scarcity", and we will approach the breeding of perennial species tolerant to abiotic stress based on studies of *Eucalyptus* under water deficiency conditions, the most common stress factor in areas used for cultivating this genus in Brazil, although stress caused by extremes of temperature, and especially cold and frosts, can be very important in the south of the country.

At present, the area planted with the *Eucalyptus* in Brazil is just over 4.5 million hectares (11.2 million acres), reflecting an average annual increase of 7.1 % during the period from 2004 to 2009. This increase occurred mainly in the states of Bahia, São Paulo, Maranhão, Mato Grosso do Sul, and Minas Gerais driven mainly by the rapid growth of *Eucalyptus* species, combined with the short rotation and high productivity of eucalyptus forests (Abraf 2010). However, in many cases this expansion of forestry activities occurred in regions with different edaphoclimatic conditions than those traditionally associated with the genus, posing a new challenge for forestry breeders: to develop high-productivity genetic material with good adaptation to these new frontiers. In these regions, lengthy periods of drought are common, resulting in low soil and atmospheric water availability which, combined with high temperatures and radiation levels, intensifies the effects of water stress. Drought or water deficiency has been pointed to as the most important stress and growth-limiting factor affecting plants. Since water deficiency is a very variable environmental condition, it is no easy task to identify and recommend genotypes

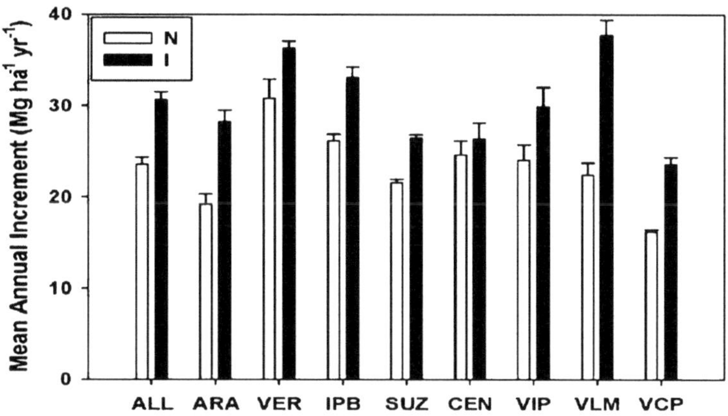

Fig. 10.1 Effect of irrigation (N, no irrigation; I, irrigation) on the mean annual increment of eucalyptus clones at different sites in Brazil (*ALL* average for all sites; *ARA* Aracruz, Espírito Santo; *VER* Eunápolis, Bahia; *IPB* Mogi Guaçu, São Paulo; *SUZ* Teixeira de Freitas, Bahia; *CEN* Guanhães, Minas Gerais; *VIP* Luís Antônio, São Paulo; *VLM* Bocaiúva, Minas Gerais; *VCP* Luís Antônio, São Paulo). *Source* STAPE et al. (2010)

tolerant to the widely varying water stress levels in these regions. Research on plant responses to water stress is therefore important for understanding the effects of water deficiency so that management practices and agricultural and forestry breeding techniques can be improved (Chaves et al. 2003).

In a network of experimental forests in Brazil, setup at eight sites to evaluate the growth of cloned eucalyptus plantations and involving adjusting the supply of nutrients and water, as well as altering the uniformity of the trees within plots, it was observed that irrigation alone increased productivity by around 30 % (Fig. 10.1), with water as the key resource in determining forest productivity (Stape et al. 2010).

Against this backdrop, it is important to find out a little more about the genus *Eucalyptus,* and especially those species that are theoretically better suited to regions with limited water availability for setting up commercially managed forests.

10.2 Germplasm and Genetic Variability

It is not our objective here to present a review of the germplasm and genetic variability of *Eucalyptus.* Detailed information is already available in a number of Brazilian and international publications (Fonseca et al. 2010; Assis and Mafia 2007; Boland et al. 2006; Brooker and Kleining 2004, 2006; Eldridge et al. 1993; Barros et al. 1990, and others). For other perennial species, whether wood-producing or not, we need to find information in the appropriate literature according to the objectives of our breeding programs and the regions in which these crops are established.

Brazil has one of the largest genetic collections of eucalyptus outside its region of origin. The genus *Eucalyptus* is divided into subgenera, with *Monocalyptus* and *Symphyomyrtus* the most important in Brazil. With the exception of *Idiogenes*, these subgenera are divided into sections. Most of the species cultivated in Brazil belong to the subgenus *Symphyomyrtus,* originating in the north of Queensland, Australia, and in particular those belonging to the section *Exsertaria* (including *E. camaldulensis, E. tereticornis*, and *E. brassiana*), which are best suited to drought conditions. These species occur naturally in three regions with average annual rainfall varying between 250 and 625 mm (*E. camaldulensis*), 1,000 and 2,500 mm (*E. brassiana*), and 600 and 2,500 mm (*E. tereticornis*) (Fonseca et al. 2010). It is only to be expected, therefore, that there should be wide variation among the subsamples within these species, and priority should be given to species from regions with lower rainfall for developing drought-tolerant genotypes. *Eucalyptus pellita* and *E. resinifera* are examples of other species that could do well under drought conditions, as well as *E. crebra* and *E. longirostrata*, two non-traditional species in Brazilian eucalyptus cultivation. In general, since drought-tolerant species exhibit low growth, plantations consisting solely of these species are of no commercial interest, but these species could be used in interspecific hybridization programs, crosses with like species or in combination with interspecific hybrids obtained from other species (Assis and Mafia 2007), provided that drought tolerance is combined with good productivity.

10.3 Inheritance, the Maternal Effect and Relations Between Characters

Very little is known about how the main characteristics related to the tolerance of abiotic factors in forestry species can be genetically controlled. Drought tolerance in particular is considered a polygenic character, making it difficult to approach using conventional genetic improvement. However, some studies of eucalyptus clones in semi-controlled conditions, subjecting seedlings to different water availability conditions, have resulted in good genetic control of physiological characteristics, such as photosynthetic rate (A), stomatal conductance (gs) and transpiration (E), with genotypic determination coefficient (H^2) values of the order of 0.5 to 0.7 (Table 10.1). These characteristics are extremely sensitive, changing at the first signs of water deficiency, and are therefore important characteristics to monitor during studies. This kind of information under field conditions is practically nonexistent.

An important observation was made during these studies: in theory, there is no evidence of the maternal effect on inheritance of the main physiological characteristics associated with water deficiency, as can be see from the data obtained during an evaluation of 166 progenies from crossing a tolerant hybrid and with a hybrid sensitive to water deficiency, under two water availability conditions, in which the average performance of the progenies was similar to that of the male parent (Table 10.2).

Table 10.1 Estimates of the genotypic determination coefficient (H^2) and genetic variance coefficient (CVg) for photosynthetic rate (*A*), stomatal conductance (*gs*), and transpiration (*E*) in eucalyptus clones subjected to different water availability conditions

	A	*gs*	*E*	Source
H^2	0.56	0.75	0.70	Revolti (2010)
	0.66	0.58	0.69	Vellini et al. (2008)
CVg (%)	6.31	13.42	8.41	Revolti (2010)
	11.20	18.36	14.44	Vellini et al. (2008)

Table 10.2 Averages for photosynthetic rate (*A*, μmol m^{-2} s^{-1}), stomatal conductance (*gs*, mol m^{-2} s^{-1}), transpiration (*E*, mmol m^{-2} s^{-1}), and leaf relative water content (RWC, %) in a segregating eucalyptus clone population subjected to two water availability conditions

Generation	*A*	*gs*	*E*	RWC
	Daily irrigation to 60 % maximum soil water retention			
Male parent (sensitive)	13.6	0.38	3.90	93.5
Female parent (tolerant)	14.9	0.45	4.22	92.6
Progenies (n=166)	13.7	0.42	3.98	92.8
	Irrigation suspended until the initial symptoms of wilting appeared, followed by irrigation to 60 % maximum soil water retention (average for four cycles)			
Male parent (sensitive)	5.0	0.09	1.27	79.8
Female parent (tolerant)	7.3	0.16	1.95	82.9
Progenies (n=166)	5.2	0.10	1.37	80.5

Source R.C. Paula, unpublished data

Morphological alterations were among the main visible changes in plants subjected to water stress. Reduction of leaf surface area, stronger root system growth, senescence, and frequently falling leaves have been reported in plants suffering from water deficiency. However, these alterations occur over the long term. The most common physiological and biochemical responses to water deficiency include closure of the stomata with a drop in stomatal conductance, rate of photosynthesis and transpiration, a drop in the photochemical efficiency of photosystem II, lower water potential and water content in the leaves, and an increase in the synthesis of osmoprotectors, such as proline and sugars.

The photosynthetic limitation brought about by water stress occurs because of both stomatal closure and the adverse effect on metabolic behavior. The fact that stomatal conductance (*gs*) is closely linked with water loss (transpiration), carbon gain, and water conservation by stomatal closure under heat stress, means that *gs* is a key parameter in evaluating the intensity of water stress. It is associated with a number of photosynthetic parameters and provides a useful tool in genetic improvement to obtain higher productivity in environments subject to drought.

Reducing stomatal conductance and transpiration, leading to a drop in the photosynthetic rate (*A*) as a result of stomatal closure are the plant's first responses in an attempt to reduce water loss under drought conditions. However, greater

Fig. 10.2 Relations between stomatal conductance, net assimilation rate (*unbroken line*), and intrinsic water use efficiency (*broken line*) in *Eucalyptus globulus* grown in the greenhouse. *Source* Pita et al. (2005)

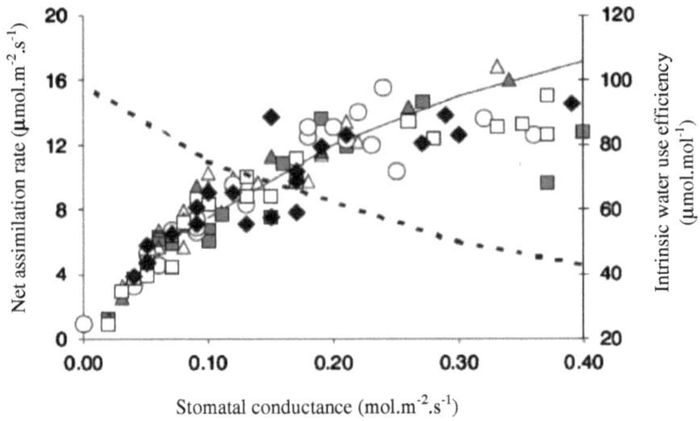

water use efficiency (WUE = *A*/*gs*) as water deficiency conditions worsen is more a question of a proportional drop in stomatal conductance values than in assimilation values (Fig. 10.2). According to Pita et al. (2005), the tendency for WUE to increase as the stomata close up often means that selecting the plants with the highest WUE is equivalent to selecting for low productivity. However, these authors go on to say that maximizing water use efficiency in the long term does not necessarily mean the continuous maximization of the WUE ratio, since plants exhibit a continuum of responses to drought, ranging from gaseous exchange, the predominant response, to morphological adjustments throughout the plant. In the long term, the expression of all these responses is what determines productivity.

Responses to water stress vary from one eucalyptus species to another. In *E. camaldulensis* and *E. saligna*, the root systems deepen, but in *E. leucoxylon* and *E. platypus,* osmotic potential drops and the cell modulus of elasticity is maximized (White et al. 2000).

In studies of six eucalyptus species, Merchant et al. (2007) observed more accentuated drops in stomatal conductance than in photosynthetic rate (Fig. 10.3). In *E. tricarpa* and *E. camaldulensis*, they observed higher values for water potential and leaf relative water content (Fig. 10.4), with less evidence of the effects of water deficiency. According to the authors, there was a drop in the accumulation of dry matter in all six species and the transpiration surface area was adjusted according to water availability. For these researchers, the simultaneous drop in osmotic potential (redistribution of cellular constituents) and total leaf surface area suggests that physiological and morphological adaptations are coordinated. It is possible that these characteristics confer some advantage on these species when grown under drought conditions.

In *E. argophloia*, there was less of a drop in the photosynthesis rate, higher biomass production, higher transpired water use efficiency (ratio of the biomass produced to the volume of water lost through transpiration), lower instantaneous water use efficiency (*A*/*gs*), and lower carbon isotope content ($\delta^{13}C$), by comparison with *E. cloeziana* (Ngugi et al. 2003a, b; 2004).

The variations observed in net assimilation rate, stomatal conductance, transpiration, water use efficiency, and leaf water potential in four eucalyptus clones under different water availability conditions identified clone 11 as the most

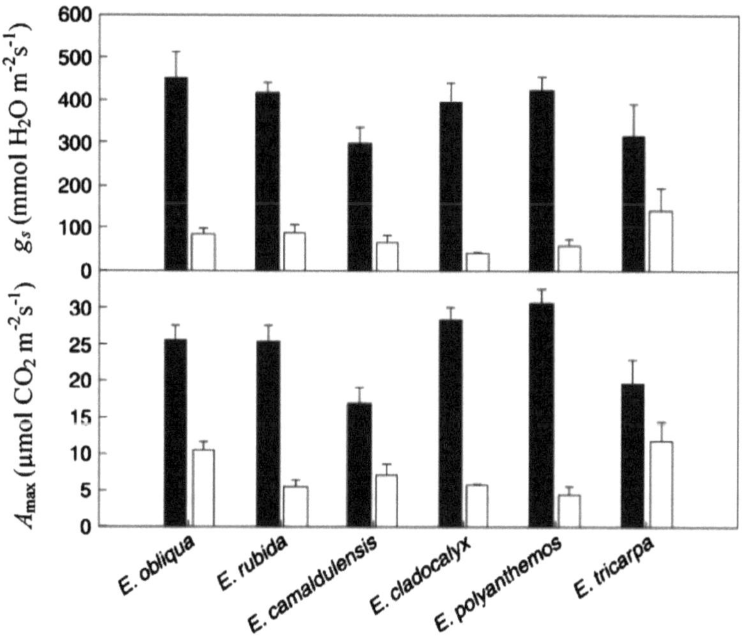

Fig. 10.3 Light-saturated photosynthesis rate (A_{max}) and light-saturated stomatal conductance ($gs_{,max}$) of six Eucalyptus species when well irrigated (field capacity—*solid bars*) and under water deficiency conditions (20 % field capacity—*open bars*). Mean and s.e. of n = 6 trees for each data point. *Source* Merchant et al. (2007)

sensitive to water stress and clones 97 and 75 as the most recommended for cropping in regions prone to drought (Table 10.3).

Species with larger and deeper root systems responded better to water deficiency, since they take advantage of a greater volume of the soil, absorbing water from deeper layers, as observed by Moroni et al. (2003) and Reis et al. (2006).

Taken together, these results show how difficult these studies can be in terms of understanding the responses of different species to water stress, indicating that there is no universal mechanism for responding to drought, but a complex pattern of responses involving morphological, anatomical, physiological, and biochemical changes. Although these changes are often evaluated in isolation, individual species use combined mechanisms for surviving and maintaining productivity under adverse conditions.

10.4 Stress Induction

Under field conditions, water stress can either build up slowly and gradually, or occur rapidly over a short time. For a gradual build up, plants can prevent dehydration by reducing their life cycles or optimizing resources as time progresses using

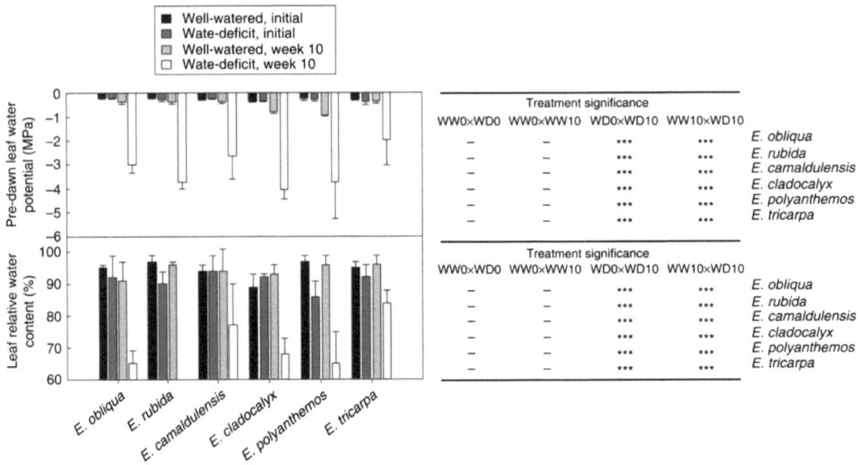

Fig. 10.4 Predawn water potential (MPa) and relative water content (%) of six Eucalyptus species for well-watered treatment (WW0) and water-deficit treatment (WD0) before commencement of treatments, (18 December, 2003) and well-watered (WW10) and water-deficit plants (WD10) at week 10 (post-water-deficit treatment, 2 March, 2004) (n = 6). Mean and s.e. are shown together with the significance of the difference between treatments and sampling dates obtained from Tukey's HSD test (*P=0.05–0.01, **P=0.01–0.001, ***P < 0.001). All trees at 0 weeks and well-watered trees at 10 weeks were watered daily to field capacity of the potting material. Plants exposed to water deficit received 20 % of the average water volume used by well-watered plants of that species. *Source* Merchant et al. (2007)

acclimation responses. For rapidly occurring water stress, plants react by minimizing water loss and/or using metabolic protection (induced or constitutive) against the harmful effects of dehydration (Fig. 10.5). In terms of physiological responses and adaptation, desiccation can produce significantly different results depending on whether dehydration is rapid or gradual, but the duration and intensity of the stress can also drastically alter the way in which plants respond, according to genotype and environmental conditions (Chaves et al. 2003).

The main effect of water stress is to reduce carbon fixation by stomatal closure, which can be initiated in plants even under moderate water stress. Plants can withstand drought conditions and avoid dehydration by keeping their water potential as high as possible or tolerating low water potential. Preventing dehydration is associated with a set of adaptive characteristics, minimizing water loss or maximizing water absorption. Water loss can be minimized in three ways: a) by closing the stomata, b) by reducing light absorption (rolling up or angling the leaves or dense layer of trichomes which increases reflectance), or c) by reducing leaf surface area (growing smaller leaves or losing old leaves). Water absorption is maximized by adjusting the pattern of allocation of assimilates, usually to increase root development (Chaves et al. 2003). According to Chaves et al., many of the characteristics that elucidate plant adaptation to drought conditions are those that determine the development and form of the plant, such as phenology, root system

Table 10.3 Photosynthetic rate (*A*), stomatal conductance (*gs*), transpiration (*E*), water use efficiency (WUE), and predawn leaf water potential (Ψh) of four eucalyptus clones subjected to three irrigation systems: irrigation to field capacity; restarted irrigation—irrigation suspended for 45 days, then irrigation to field capacity for a further 55 days; and water deficiency—irrigation suspended for 100 days

Irrigation management for each clone	*A*	*gs*	*E*	WUE	Ψh
			Clone 97		
Irrigation to field capacity	12.02 a	0.305 a	7.64 a	1.58 a	−1.13 b
Restarted irrigation	10.95 a	0.395 a	5.92 a	1.86 a	−1.16 b
Water deficiency	5.82 b	0.265 a	6.47 a	0.91 b	−2.60 a
			Clone 11		
Irrigation to field capacity	9.66 a	0.147 a	6.57 a	1.49 a	−1.83 b
Restarted irrigation	8.56 a	0.157 a	5.84 a	1.50 a	−1.86 b
Water deficiency	1.19 b	0.017 b	1.40 b	0.91 b	−2.80 a
			Clone 75		
Irrigation to field capacity	10.93 a	0.242 a	6.41 a	1.72 a	−1.03 b
Restarted irrigation	9.74 a	0.232 a	6.92 a	1.43 a	−1.06 b
Water deficiency	7.18 b	0.225 a	5.90 a	1.20 a	−3.16 a
			Clone 84		
Irrigation to field capacity	9.62 a	0.235 a	5.95 a	1.68 a	−1.03 b
Restarted irrigation	7.42 ab	0.115 a	4.76 a	1.62 a	−1.00 b
Water deficiency	5.39 b	0.182 a	6.31 a	0.89 b	−4.70 a

Averages followed by the same letter in the column for each clone did not differ in the Tukey test at 5 % probability. *Source* Tatagiba et al. (2008)

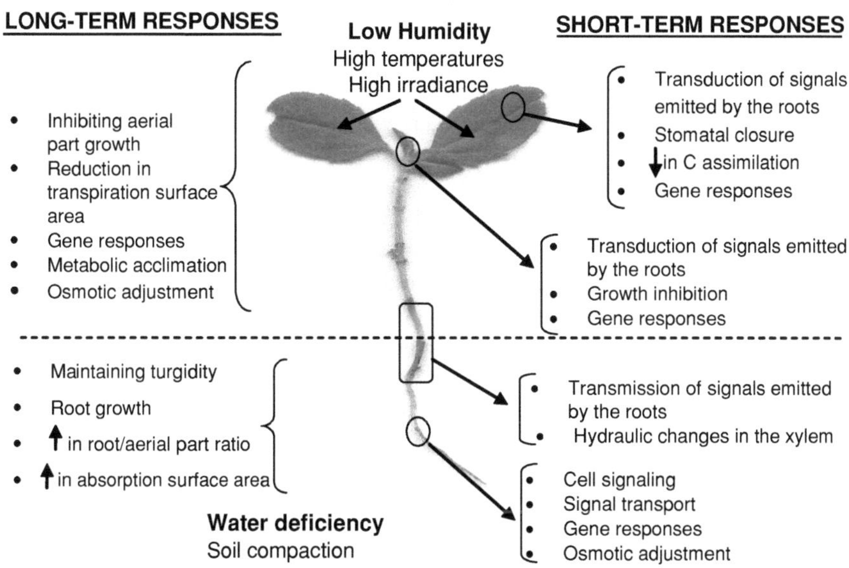

Fig. 10.5 Plant responses to water stress. *Left*: long-term responses or acclimation; *Right*: short-term responses. *Source* adapted from Chaves et al. (2003). Authorized by CSIRO and CSIRO PUBLISHING (http://www.publish.csiro.au/nid/102/paper/FP02076.htm)

size and depth, the properties of the xylem or storage of reserves. Furthermore, these characteristics are constitutive and not stress induced.

Water stress also induces a number of biochemical and physiological responses, such as the build up of sugars (sucrose, trehalose, sorbitol and raffinose family oligosaccharides), mannitol, amino acids (proline), and amines (glycine betaine and polyamines). Water deficiency induces the production of abscisic acid (ABA), causing stomatal closure and inducing the expression of genes related to water stress.

Low water potentials can be tolerated by making osmotic adjustments, increasing cell wall rigidity, or simply producing smaller cells. Many arboreal species in arid and semi-arid regions combine high solute concentration (low osmotic potential) with sclerophylly, low photosynthetic capacity, and stomatal conductance. However, this does not appear to be the case with some species of *Eucalyptus* and *Banksia* that have scleromorphic leaves and exhibit a high maximum photosynthesis rate per unit area (Chaves et al. 2003).

Tolerance to water stress varies from one genotype to another, requiring the adjustment of metabolic pathways in a process known as acclimation. Initial stress responses are aimed at immediate survival, whereas acclimation (metabolic and structural change resulting from alterations in gene expression) helps to improve the plant's performance and functioning under stress conditions (Bohnert and Sheveleva 1998). The low growth commonly observed in drought-tolerant plants is associated with the metabolic cost involved in the majority of water stress responses. For instance, stomatal closure reduces carbon assimilation; increased root growth is achieved at the expense of aerial part growth; discarding leaves entails the cost of producing new leaves and the build up of solutes in the cell alters the metabolism of enzymes and proteins (Taiz and Zeiger 2004).

Unlike some annual plants, arboreal species are not able to escape drought and have therefore developed many mechanisms to withstand it. Bearing in mind that each type of strategy is species-specific, knowledge of these strategies allows us to select physiological characteristics that can be used in breeding programs.

Studying tolerance to an abiotic factor in perennial species under field conditions is not always easy. This type of study presents a number of limitations, such as enormous cropping areas, common in experiments with these crops, uncertainty as to whether the stress event (drought, cold, high temperatures, and others) will occur during the crop cycle, the occurrence of the event during phenological or critical development stages, as well as the actual intensity and duration of the stress factor. In other words, the magnitude of effects of different types of abiotic stress depends on a series of factors, such as intensity and duration, the plant's development stage, and its genetic capability to respond to these stress factors.

Since working under controlled or semi-controlled conditions poses problems in the field, especially since the majority of arboreal crops are extremely tall, work on drought tolerance is often carried out indirectly or at the seedling stage. This is a key stage in the life cycle of arboreal species, since survival capability is dependent on all previous stages. Seedlings have fairly undeveloped root systems that limit access to soil water, which means that they are especially vulnerable to water stress. Perhaps this is why many studies on arboreal species refer to the seedling as the initial growth phase.

Another critical phase is the closure of the canopy at 2–4 years growth, during which competition among plants for environmental resources becomes more intense. This is why questions are often raised as to whether the results obtained from potted seedlings, simulating the forest layout (from sowing to around 90 or 120 days), are actually representative of what the plant can withstand during its life cycle in the forest. In fact, this question is not easy to answer. For instance, we know that some species exhibit striking differences between the juvenile and adult stages in regard to their aerial and root growth habits. In terms of leaves, some species exhibit changes in leaf morphology and arrangement, and in the production of waxes. According to Grattapaglia (2007), during their life cycles, trees are often subjected to variable environmental conditions, a feature that has contributed significantly to phenotypic variance. This author has also stated that morpho-physiological changes in trees during the life cycle suggest that different sets of genes are involved, or the same genes with effects of different magnitude, making it difficult to detect the corresponding QTLs (*Quantitative Trait Loci*).

The majority of water stress experiments on forestry species simulate the stress by suspending irrigation altogether, by varying the soil water content, or by inserting dry periods into the irrigation cycle. Each of these procedures has advantages and disadvantages, and the characteristics evaluated must be carefully planned and defined to adequately simulate the phenomenon to be measured. For example, when irrigation is suspended altogether, growth characteristics may be of secondary importance, whereas physiological and/or biochemical changes are more marked and therefore more appropriate for measurement purposes. Under field conditions, stress is usually simulated by covering over the soil and channeling rainwater in order to intercept it and vary its intensity. Whatever the experimental conditions, whether in the field or in the greenhouse, phenotyping must be carried out to guarantee the reliability of the results and recommend breeding and management practices.

Another aspect to be taken into account in species grown for timber production and not for fruit or seeds is that the phenological stage at which the stress event occurs may not be important in the way that stress before budding, florescence, or even fruit development is. However, if the aim is fruit or seed production, or to provide support for breeding programs, the occurrence of water stress in these crucial phases can compromise the entire recombination process for the year, and therefore ultimately the seed yield itself.

Although perennial species are more subject to adverse factors during their life cycle, they also live longer and can therefore take advantage of years in which there is abundant rainfall to offset the loss of forest growth caused by a severe drought. In other types of perennial species, where the product of interest is obtained annually, such as fruit species, within certain limits the losses in bad years are no different from species with shorter cycles, and can compromise the year's production. The fact is that during the development of a breeding program, stress factors may not occur at all or occur at lower intensity, so that there is no guarantee that the materials selected for continuing the breeding program will survive if subjected to more intense stress events.

10.5 Strategies and Methods for Selection and Breeding

The forestry breeding programs conducted in Brazil have been primarily directed toward growth characteristics, and more recently, timber quality. Indirectly, there has been selection, albeit unintentional, for adaptation to the environmental conditions under which commercial companies operate, and it can be assumed that drought tolerance characteristics have not been fully explored in these programs.

As new forest frontiers have opened up in regions with limited water resources, the search for productive, and therefore drought-tolerant genotypes should be a key feature in breeding programs. Since the species concerned are perennials with a long life cycle, during their development the plants pass through innumerable situations that are unfavorable to growth. In contrast to short-cycle species, trees cannot avoid these conditions. This means that the mechanisms for withstanding abiotic stress factors in perennial species may be more complex and primarily aimed at species survival, to the detriment of productivity. It has already been observed that drought-tolerant genotypes are of low productivity, since drought tolerance mechanisms involve metabolic costs, such as stomatal closure, production of osmoprotective compounds, deepening root systems to the detriment of the aerial part, etc. In view of these features, drought-tolerant species are not recommended for large-scale commercial plantations. For this reason, the most promising breeding strategy for achieving abiotic stress tolerance in eucalyptus is hybridization.

Interspecific hybridization aims at achieving a combination of good technical characteristics for the timber, tolerance to biotic and abiotic stresses, and heterosis, as verified in a number of hybrid crosses (Assis and Mafia 2007). According Assis and Mafia, crossing productive but drought-sensitive species with drought-tolerant but low-productivity species has produced individuals with higher productivity than their tolerant parents.

Once favorable hybrid combinations have been obtained, superior individuals can fairly easily be multiplied by vegetative propagation. Thus, cloning combined with hybridization provides an appropriate way of fixing characteristics of interest and guaranteeing plantation uniformity with perceivable gains throughout the production chain.

Detailed descriptions of eucalyptus breeding methods and strategies are given in Fonseca et al. (2010); Resende and Higa (1990) and other authors.

10.6 Biotechnology Applied to Breeding Perennial Species for Abiotic Stress Conditions

High-performance genomic technologies integrated into forestry breeding programs are opening up new perspectives for gene and genome study, helping us to understand the complex relations between genetic variability and phenotypic diversity.

Even under ideal edapho climatic conditions for planting eucalyptus in Brazil, leading to rapid growth, the long selection cycle remains a limiting factor in the development of forestry breeding programs. The general trend in forestry breeding is toward integrating conventional methods and modern biotechnology, taking into account the advantages and limitations of both. These new methods can make a significant contribution to our basic knowledge of the crop and the character under investigation, and also help in developing and producing improved clones.

Recently, the development of standard routines for cloning, sequencing, genetic expression assessment, and statistical and computational analysis have lead to unprecedented advances in the genetic study of various organisms. The use of biotechnology as a tool for breeding improved forestry species is an important alternative. Using this technology effectively in breeding forestry species, with the aim of making early selection more efficient and thereby speeding up and increasing genetic gains, presents the best prospects for success.

Initial studies based on the use of molecular markers in forestry species showed that it was possible to use simple pedigrees, such as families and half and full siblings, to identify links between markers and characteristics of interest. Incorporating these methods into forestry species breeding is a slow ongoing process that will continue as these species are domesticated and improved.

Over the last few years, based on the potential use of the information produced by the molecular biotechnology tools available, eucalyptus breeding programs have been set up to obtain different segregating progenies for a number of characteristics of interest in forestry production. Furthermore, advances in molecular biology promise to improve our understanding of how variations in the DNA sequence affect the phenotype. Based on this knowledge, it will be possible to select those sequences favorable to gains in characters of interest by means of molecular improvement.

Molecular biology and biotechnology, together with conventional forestry species breeding, could help cut the time required to complete each selection and recombination cycle, a great advantage when dealing with perennial species. Early marker-assisted selection is useful for improving characters related to timber quality, which is generally difficult to measure, requiring expensive, laborious procedures that can only be conducted when the timber is advanced in age.

There are two general strategies that can be used to identify the genes controlling qualitative and quantitative phenotypic variation: reverse genetics and forward genetics. Reverse genetics concentrates on the gene sequence, analyzing natural or induced (mutagenic) variations in order to find a link with a phenotypic or physiological change. Reverse genetic approaches include studies of directed mutagenesis, gene silencing, and super expression. Forward genetics entails analyzing the phenotype and locating genome regions containing genes that control part of the characteristic's variability. This strategy is based on gene mapping and positional cloning.

Reverse genetics is partly based on a knowledge of the full or partial gene sequence of the organism under investigation. With the prospect of applying this procedure and bearing in mind that few gene sequences of the genus *Eucalyptus* have been deposited in public databases, the Brazilian paper and cellulose companies, together with research incentive agencies such as the Brazilian National Council for Scientific and Technological Development (CNPq), São Paulo Research Foundation (FAPESP), and Brazilian Study and Project Financing Agency (FINEP) have launched initiatives to develop two independent projects, *Genolyptus* and *Forests*, with the aim of setting up databases of expressed sequence tags (ESTs) in eucalyptus.

One strategy for identifying regions of the genome-containing control genes or sequences involved in determining characters of economic importance is gene mapping associated with quantitative analysis and the use of bulk segregant analysis (BSA) of quantitative characteristics. These forward genetics strategies are advantageous because no prior knowledge of the candidate gene sequence is required.

Genetic maps are used to define the statistical probability that certain genome regions control certain aspects of phenotypic variability. These regions are delimited by molecular markers, known as QTL. The markers associated with the QTLs can be used in marker-assisted selection or as guides for positional cloning of genes or regions involved in controlling the character.

Although very important, genetic mapping in perennial species is not an easy task. Calculating the recombination frequency for pairs of loci is more complex in *Eucalyptus* than in annual species for which endogamic lines can be obtained. With the pseudo-testcross strategy (Grattapaglia and Sederoff 1994), in which segregant markers are analyzed separately in each parent, it is possible to use heterozygotic parents in constructing linkage maps. A number of groups identify genome regions associated with characters of interest in *Eucalyptus*, including height, leaf size, seedling tolerance to frost, vegetative propagation capability, disease resistance, growth and timber quality, early florescence, formation of lignotubers, and seedling abnormalities.

This work allows us to estimate the minimum number of loci involved in controlling the characters, the genome locations of these loci, and how much they contribute to phenotypic variation. We can also assess their effects in different environments and genotypes, substantially improving our understanding of the architecture of these complex characters. However, this work is limited by the small size of the progenies used, by restricted phenotyping precision, by the small number of crosses tested, and by the limitations inherent in the molecular marker types, pedigrees, and statistical methods used. This means that we do not know enough about the actual number of QTLs, their precise positions, and the magnitude of their effects, so that, despite all the effort that goes into studying complex characters, we still know very little about their molecular bases.

Although it is easy to obtain large progenies and replicate many species, only a small number of control genes or sequences have been identified as responsible for some of the variation in these characters. In spite of the numerous studies on QTL mapping, not just in *Eucalyptus*, but also in *Populus* and *Pinus*, as yet it has not been possible to confirm that a gene is responsible for any QTL. A further limitation is the fact that most of these studies involved using dominant markers that are not normally transferable to other crosses, limiting inter-experimental data sharing and the possible use of markers linked to the QTL in other crosses involving marker-assisted selection.

Our knowledge of the structure of the eucalyptus genome is growing rapidly due to the initiative of various research groups concerned with providing information relevant to forestry breeding, and not just academic knowledge. The information that will be made available by the recently finalized work to completely sequence the eucalyptus genome will also produce a large volume of information on the organization of the genome, together with an enormous quantity of markers obtained from the DNA.

The increasing availability of partial or complete gene sequences from forestry genome projects, as well as complete model plant genomes, has allowed us to develop markers based on gene sequence variability.

In theory, the use of genes as references for selecting trees is the best choice, since it will enable us to establish direct links between the variability of these gene sequences and the phenotyping observed. The great advantage of taking this approach, specifically in trees, is that the potentially limiting question of gametic linkage equilibrium between molecular marker alleles and the linked gene alleles becomes irrelevant. Furthermore, this approach enables us to analyze germplasm banks and collections of elite clones characterized in detail, eliminating the time required for producing, planting, and measuring segregating populations.

High-throughput genotyping techniques are already available to produce a significant number of candidate gene analyses very rapidly.

In future, the integration of genomic strategies for discovering structural and expressive allele variability with high marker-density gene mapping, especially mapping elite populations and using transgenics for function testing, will provide great opportunities for elucidating wood formation processes and biotic and abiotic stress responses in trees.

References

Abraf–Associação Brasileira de Produtores de Florestas Plantadas (2010) Anuário estatístico da ABRAF 2010 ano base 2009. http://www.abraflor.org.br/estatisticas/ABRAF10-BR.pdf. Acessed 06 April 2011

Assis TF, Mafia RG (2007) Hibridação e clonagem. In: Borém A (ed) Biotecnologia florestal. Suprema Gráfica e Editora Ltda, Viçosa

Barros NF, Novais RF, Cardoso JR, Macedo PRO (1990) Algumas relações solo-espécies de eucalipto em suas condições naturais. In: Barros NF, Novais RF (eds) Relação solo-eucalipto. Editora Folha de Viçosa, Viçosa

Bohnert HJ, Sheveleva E (1998) Plant stress adaptations–metabolism move. Curr Opin Plant Biol 1:267–274. doi:10.1016/S1369-5266(98)80115

Boland DJ, Brooker MIH, Chippendale GM, Hall N, Hyland BPM, Johnson RD, Kleining DA, McDonald MW, Turner JD (2006) Forest trees of Australia, 5th edn. Collingwood VIC, CSIRO Publising

Brooker MHI, Kleining DA (2004) Field guide to eucalypts, 2nd edn., vol 3. Boomings books, Melbourne

Brooker MHI, Kleining DA (2006) Field guide to eucalypts, 3rd edn., vol 1. Boomings books, Melbourne

Chaves MM, Maroco JP, Pereira JS (2003) Understanding plant responses to drought–from genes to the whole plant. Funct Plant Biol 30:239–264. doi:10.1071/FP02076 (http://www.publish.csiro.au/nid/102/paper/FP02076.htm)

Eldridge K, Davidson J, Harwood C, van Wyk G (1993) Eucalypt domestication and breeding. Clarendon, Oxford

Fonseca SM, Resende MDV, Alfenas AC, Guimarães LMS, Assis TF, Grattapaglia D (2010) Manual prático de melhoramento genético do eucalipto. UFV, Viçosa

Grattapaglia D (2007) Mapas genéticos e seleção assistida por marcadores moleculares. In: Borém A (ed) Biotecnologia florestal. Suprema Gráfica e Editora Ltda, Viçosa

Grattapaglia D, Sederoff R (1994) Genetic linkage maps of *Eucalyptus grandis* and *Eucalyptus urophylla* using a pseudo-testcross: mapping strategy and RAPD markers. Genetics 137: 1121–1137

Merchant A, Callister A, Arndt S, Tausz M, Adams M (2007) Contrasting physiological responses of six *Eucalyptus* species to water deficit. Ann Bot 100:1507–1515. doi:10.1093/aob/mcm234

Moroni MT, Worledge D, Beadle CL (2003) Root distribution of *Eucalyptus nitens* and *E. globulus* in irrigated and droughted soil. For Ecol Manage 177:399–407. doi:10.1016/S0378-1127(02)00410-3

Ngugi MR, Doley D, Hunt MA, Ryan P, Dart P (2004) Physiological responses to water stress in *Eucalyptus cloeziana* and *E. argophloia* seedlings. Trees 18:381–389. doi:10.1007/s00468-003-0316-5

Ngugi MR, Hunt MA, Doley D, Ryan P, Dart P (2003a) Dry matter production and allocation in *Eucalyptus cloeziana* and *Eucalyptus argophloia* seedlings in response to soil water deficits. New Forest 26:187–200. doi:10.1023/A:1024493917483

Ngugi MR, Hunt MA, Doley D, Ryan P, Dart P (2003b) Effects of soil water availability on water use efficiency of *Eucalyptus cloeziana* and *Eucalyptus argophloia* plants. Aust J Bot 51: 159–166. doi:10.1071/BT02103

Pita P, Cañas I, Soria F, Ruiz F, Toval G (2005) Use of physiological traits in tree breeding for improved yield in drought-prone environments. The case of *Eucalyptus globulus*. Invest Agr: Sist Recur For 14:383–393

Reis GG, Reis MGF, Fontan ICI, Monte MA, Gomes NA, Oliveira CHR (2006) Crescimento de raízes e da parte aérea de clones de híbridos de *Eucalyptus grandis* x *Eucalyptus urophylla* e de *Eucalyptus camaldulensis* x *Eucalyptus* spp. submetidos a dois regimes de irrigação no campo. Revista Árvore 30(6):921–931. doi:10.1590/S0100-67622006000600007

Resende MDV, Higa AR (1990) Estratégias de melhoramento para eucaliptos visando a seleção de híbridos. Bol Pesq Fl 21:49–60

Revolti PM (2010) Divergência genética entre clones de eucalipto por caracteres biométricos e fisiológicos sob deficiência hídrica. Jaboticabal: FCAV/UNESP, 2010. p 66 (Master Thesis)

Stape JL, Binkley D, Ryan MG, Fonseca S, Loos RA, Takahashi EM, Silva CR, Silva SR, Hakamada RE, Ferreira JMA, Lima AMN, Gava JL, Leite FP, Andrade HB, Alves JM, Silva GGC, Azevedo MR (2010) The Brazil Eucalyptus potential productivity project: influence of water, nutrients and stand uniformity on wood production. For Ecol Manage 259:1684–1694. doi:10.1016/j.foreco.2010.01.012

Taiz L, Zeiger E (2004) Fisiologia vegetal, 3rd edn. Artmed, Porto Alegre

Tatagiba SD, Pezzopane JEM, Reis EF (2008) Relações hídricas e trocas gasosas na seleção precoce de clones de eucalipto para ambientes com diferenciada disponibilidade de água no solo. Floresta 38(2):387–400

Vellini ALTT, Paula NF, Alves PLCA, Pavani LC, Bonine CAV, Scarpinati EA, Paula RC (2008) Respostas fisiológicas de diferentes clones de eucalipto sob diferentes regimes de irrigação. Revista Árvore 32(4):651–663. doi:10.1590/S0100-67622008000400006

White DA, Turner NC, Galbraith JH (2000) Leaf water relations and stomatal behavior of four allopatric *Eucalyptus* species planted in Mediterranean southwestern Australia. Tree Physiol 20:1157–1165. doi:10.1093/treephys/20.17.1157

Index

A
Abiotic stress, 94, 95, 99, 100, 123, 137, 145, 151, 152, 157–159, 161, 163, 165–169
Abscissic acid (ABA), 166
Absorption of nutrients, 103
Acclimation, 164–166
Acid phosphatase, 39
Acid soils, 123, 124, 129, 133–35
Activation state, 32
Al complexation, 41
Al exclusion, 43
Al immobilization, 43
Al tolerance, 41
Al toxicity, 41
Alternative metabolic pathways, 40
Alternative oxidase pathway, 40
Alternative respiratory systems, 39
Aluminum toxicity, 123, 124, 134–136
Antioxidant system, 25, 27, 29, 43

B
Biochemical responses, 161
Biological N fixation, 37
Biotechnology, 88, 98, 102, 168, 169
Brazil, 157–160, 168
Breeding methods for salinity, 103
Breeding programs, 157, 159, 166–168
Bulk segregant analysis (bsa), 170

C
C4 metabolism, 25
Canopy architecture, 33
Canopy extinction coefficient, 36
Carbon gain, 32
Carbon isotope, 162
Carbon isotopic discrimination, 24
Cation exchange capacity, 42
Cell division, 42
Cell expansion and elongation, 42
Cell modulus of elasticity, 162
Chelating or reducing agents, 40
Cloning, 168–170
Compatible osmolytes, 28, 32
Crop breeding, 137

D
Dehydration, 163, 164
Desiccation, 164
DNA, 169, 171
Drought, 22, 157, 158, 160–164, 166–168
Drought tolerance, 26
Drought-sensitive, 168
Drought-tolerant, 157, 160, 166, 168
Dry matter or grain yield efficiency index, 103

E
E. argophloia, 162
E. brassiana, 160
E. camaldulensis, 162
E. cloeziana, 162
E. crebra, 160
E. leucoxylon, 162
E. longirostrata, 160
E. platypus, 162
E. resinifera, 160
E. saligna, 162
E. tereticornis, 160
E. tricarpa, 162

R. Fritsche-Neto and A. Borém (eds.), *Plant Breeding for Abiotic Stress Tolerance*,
DOI: 10.1007/978-3-642-30553-5, © Springer-Verlag Berlin Heidelberg 2012

E (*cont.*)
Early selection, 169
Electrical conductivity, 103, 106–109, 112, 119
Eucalyptus breeding, 168, 169
Eucalyptus genome, 171
Eucalyptus, 158–172
Expressed sequence tags (ests), 170
Exudation of organic acids, 38, 43

F
Flavonoids, 40
Floral abortion, 31
Forestry breeding, 157, 159, 168, 169, 171
Forward genetics, 169, 170

G
Gene mapping, 169, 171
Gene, 158–161, 165–172
Genetic control, 160
Genetic expression, 169
Genetic gains, 169
Genetic markers, 103, 115
Genetic variability, 107, 114, 130, 159, 168
Genolyptus, 170
Genome, 168–171
Genomic, 171
Germplasm, 159–171
Global climate changes, 22
Glutamine synthetase, 35
Glycine betaine, 166
Glycophytes, 27
Green revolution, 22

H
Halophyte, 27
Heat stress, 30
Heat tolerance, 33, 137, 143, 153–156
Heterosis, 168
High-affinity transporter systems, 34
Hybridization, 160, 168
Hybrids, 160

I
Improved crops, 88
Ionic imbalance, 28

J
Jasmonic acid, 42

L
Leaf abscission, 25
Leaf senescence, 31
Leaf water potential, 162, 165
Lignin synthesis, 42
Low-affinity transporter systems, 34

M
Maintenance respiration, 24
Mannitol, 166
Marker-assisted selection, 169, 170
Maternal effect, 160
Maximum quantum yield of photosystem II, 44
Membrane permeability, 31
Mesophyll conductance, 25
Metabolic protection, 164
Mitochondrial electron transport, 40
Mitochondrial respiration, 29
Molecular biology, 169
Molecular biotechnology, 169
Molecular markers, 169, 170
Morphological adaptations, 162

N
N assimilation efficiency, 35
N deficiency, 33
N remobilization efficiency, 35
N storage capacity, 35
N uptake efficiency, 33
N uptake post-anthesis, 37
N use efficiency, 33
N utilization efficiency, 33
Nitrate reductase, 35
Nitrate reduction, 36

O
Oligosaccharides, 166
Osmoprotectors, 161
Osmotic adjustment, 25, 28, 166
Osmotic effect, 28
Oxidative phosphorylation, 40
Oxidative stress, 28

P
P acquisition or uptake efficiency, 38
P deficiency, 39
P use efficiency, 38
P utilization efficiency, 38
Phenolic compounds, 40

Phenological stage, 167
Phenotypic variance, 167
Phenotyping, 167, 170, 171
Phenylpropanoids, 42
Phosphatase activity, 39
Photorespiration, 24, 36
Photosynthesis, 161–163, 165
Photosynthetic capacity, 24
Photosynthetic N use efficiency, 36
Photosynthetic rate (a), 160, 161, 165
Photosystem II, 163
Physiological responses, 164, 166, 172
Pi transporters, 39
Pinus, 170
Polyamines, 166
Populus, 170
Post-anthesis N absorption, 35
Post-anthesis N uptake, 34
Post-translational control, 35
Productivity, 101
Proline, 161, 166
Protoplasmic tolerance, 23
Pseudo-testcross strategy, 170

Q
Quantitative trait loci (QLT), 103, 115

R
Radiation use efficiency, 36
Raffinose, 166
Relative water content, 161, 162, 164
Reverse genetics, 169
Root architecture, 38
Root growth, 165–167
Root length density, 34
Rubisco, 25, 32
RuBP regeneration, 32

S
Salinity threshold, 103, 107, 108
Salinity tolerance, 28
Salinity, 28
Salt tolerance, 29
Sclerophylly, 40
Secondary metabolites, 40
Selection strategies, 88, 97, 137, 147, 154
Senescence, 37
Sensible heat, 30
Sequencing, 169
Signaling pathway, 28
Sorbitol, 166
Specific leaf mass, 25
Stay-green, 36
Stomatal closure, 161, 164–166
Stomatal conductance (gs), 160, 161, 163
Stomatal conductance, 23
Stress, 157–171
Sucrose, 166

T
Thermal tolerance, 31
Tolerance mechanism, 137, 140
Tolerance, 157, 158, 166
Transgenic method, 103, 116
Transgenics, 171
Transpiration (e), 160, 161, 165
Trehalose, 166
Tropical environments, 22

V
Vegetative propagation, 168, 170

W
Water availability, 158–163
Water deficiency, 88–99, 157, 158, 160, 161
Water stress, 161–168
Water use efficiency, 22, 162, 165

MIX
Papier aus verantwortungsvollen Quellen
Paper from responsible sources
FSC® C105338
www.fsc.org

If you have any concerns about our products,
you can contact us on
ProductSafety@springernature.com

In case Publisher is established outside the EU,
the EU authorized representative is:
Springer Nature Customer Service Center GmbH
Europaplatz 3, 69115 Heidelberg, Germany

Printed by Libri Plureos GmbH
in Hamburg, Germany